The Superior Project Organization

CENTER FOR BUSINESS PRACTICES

Editor

James S. Pennypacker

Director
Center for Business Practices
West Chester, Pennsylvania

ADDITIONAL VOLUMES IN PREPARATION

The Superior Project Organization
Global Competency Standards and Best Practices

Frank Toney

Executive Initiative Institute
Scottsdale, Arizona

Center for
Business
Practices

CAPTURING, ORGANIZING, AND
TRANSFERRING MANAGEMENT KNOWLEDGE

MARCEL DEKKER, INC.   NEW YORK · BASEL

ISBN: 0-8247-0638-2

This book is printed on acid-free paper.

Headquarters
Marcel Dekker, Inc.
270 Madison Avenue, New York, NY 10016
tel: 212-696-9000; fax: 212-685-4540

Eastern Hemisphere Distribution
Marcel Dekker AG
Hutgasse 4, Postfach 812, CH-4001 Basel, Switzerland
tel: 41-61-261-8482; fax: 41-61-261-8896

World Wide Web
http://www.dekker.com

The publisher offers discounts on this book when ordered in bulk quantities. For more information, write to Special Sales/Professional Marketing at the headquarters address above.

Current printing (last digit):
10 9 8 7 6 5 4 3 2 1

PRINTED IN THE UNITED STATES OF AMERICA

Series Introduction

The organizational environment needed for project success is ultimately created by management. The way that the managers define, structure, and act toward projects is critical to the success or failure of those projects, and consequently the success or failure of the organization. An effective project management culture is essential for effective project management.

This Center for Business Practices series of books is designed to help you develop an effective project management culture in your organization. The series presents the best thinking of some of the world's leading project management professionals, who identify a broad spectrum of best practices for you to consider and then to implement in your own organizations. Written with the working practitioner in mind, the series provides "must have" information on the knowledge, skills, tools, and techniques used in superior project management organizations.

A culture is a shared set of beliefs, values, and expectations. This culture is embodied in your organization's policies, practices, procedures, and routines. Effective cultural change occurs and will be sustained only by altering (or in some cases creating) these everyday policies, practices, procedures, and routines in order to impact the beliefs and values that guide employee actions. We can affect the culture by changing the work climate, by establishing and implementing project management methodology, by training to that methodology, and by reinforcing and rewarding the changed behavior that results. The Center for Business Practices series focuses on helping you accomplish that cultural change.

Having an effective project management culture involves more than implementing the science of project management, however—it involves the art of applying project management skill. It also involves the organizational changes that truly integrate this management philosophy. These changes are sometimes structural, but they always involve a new approach to managing a business: projects are a natural outgrowth of the organization's mission. They are the way in which the organization puts in place the processes that carry out the mission. They are the way in which changes will be effected that enable the organization to effectively compete in the marketplace.

We hope this Center for Business Practices series will help you and your organization excel in today's rapidly changing business world.

James S. Pennypacker

Preface

Large organizations find that implementation of professional project organizations generates sizeable amounts of incremental revenue and reduced costs. Over nearly 10 years, the experiences of the more than 80 companies that participate in the Top 500 Project Management Benchmarking Forum in starting and building project organizations have been documented. Compiled from this experience is a listing of best practices and competencies that lead to organizational success. The conclusions are supported by numerous outside research studies.

The best practices and competency standards can be immediately applied in organizations to implement project organizations and improve the effectiveness of those already in operation. Improved is capability for project organizations and offices to implement the strategy of the host organization as well as to support project teams in their efforts to attain project goals through speed, efficiency, and effectiveness.

Specific organizational performance-improving topics covered include methods of obtaining senior-level support, effective selection and nurturing of project managers, and application of speed-, efficiency-, and effectiveness-based project methodologies. Project portfolio management techniques and practices are detailed. The impact on project success of political issues and practices such as ambassadorship, impression management, public relations, promotion of and lobbying for project interests and needs is covered. Other topics include competency-based training and the application of structured, predictable, and consistent methodologies and tools. The benefits and practices of continually improving project management performance by transferring knowledge from one project to the next and implementing end-to-end project involvement by the project group are also addressed.

Frank Toney

Acknowledgments

A research effort of this magnitude necessitates the involvement of many participants and contributors—all of whom provided time on a volunteer basis. To date, the work has spanned a period of five years. The author sincerely thanks the following people and the companies they represent for their contributions in the benchmarking forums and the developments of this document.

Aeroquip, Yvette Burton
Alcoa, Alan Kristynik
Allied Signal, Tom Booth
American Airlines, Susan Garcia
AMEX, Mary Burgger
Amway/Access, Ed VanEssendelft
Arizona Republic, Larry Lytle
Avnet, Julee Rosen
Battelle, Steve S. Eschlahta
Bellsouth, Ed Prieto
Cablevision, Dave Steinbuck, Steve Potter, Cliff Hagen, Jackie Ernst

Calpers, Doug McKeever, Michael Ogata, Kristen Sawchuck
Capital One, Bob Stanley, Debbie Adams
Caterpillar Inc., Dave Harrison
Center for Business Practices, Jim Pennypacker
CH2M Hill, Starr Dehn
Chase Manhattan Bank, Ruth Guerroro
Citibank, Warren Marquis, Lou Rivera
Compass Telecom, Ray Powers

Computer Horizons, Jack Routh
Deluxe Checks, Clark Hussey
Development Dimensions International, Pat Smith
Dinsmore Associates, Paul Dinsmore
Disneyland, Bob Teal
Dow Chemical, Bill Lehrmann
DuPont Agricultural Products, Julie Eble
Eastman Kodak, Fred Bassette
EDS, Carl Isenberg, Mike Wall, Paul Schwartz
Electric Lightwave, Michael Golob
Eli Lilly, Sheldon Ort
Ernst & Young, Steve Sawle, Dan Roth
Federal Express, Don Colvin, Dinah Allison
Fortis, Inc., David Cupps
General Services Administration, John Bland, Hugh Colosacco
GlaxoSmithKline, Derek Ross
Harding Lawson Associates, Don Campbell
IBM, Sue Guthrie
InFocus Systems, Scott Stevenson
ITT Hartford, Sue Steinkemp
JD Edwards, Kelly McCormick, Nancy Tilson
Kelly Services, Pat Donahue, Joanne Bolas, Kathleen Tabaczynski, Gloria Sirosky, Deborah Polley
Kemper, Debbie Hoyt
Logistics Management, Tripp Home
MetLife, Carol Rauh
Micro Age Inc., Dan McFeely

Miller Brewing, Lowell Skelton
Morgan Stanley, Tom Tarnow, Rolan Armove, Andrew Posen
Motorola, Martin O'Sullivan, Richard Gale, Boyd Mathes
NCR, Pat Peters
Nissan Motor Corp., Randal Macdonald
Northrop Grumman, Frank Catalfamo
Northwestern Mutual Life, Marge Combe, Judy Koening, Margie Dougherty, Patricia Weber
Nynex, Kathy Kuzman
Oracle, Meg Trobiano
Philip Morris, Barbara Miller, Bob Riley
Rust Environmental, Gary Scherbert
Sabre Group, Chris Thomas, Ed Fox
SBC/AIT, Terri Hart-Sears, Margaret Currie
Southern Company, Andy Hidle, Nancy Bradley
Sprint, Kenneth Binnings, Rachel Kussman, Don Albers, Robert Ramos
SRM/Hennepin County, Derek Mantel
Texaco, Richard Legler
Transwestern Publishing, Julie Freeman
Washington Gov., Richard Humphry
West Group, Jeannene Dorle
Zurich, Joe Lunn, Jerry Egger, Brent Kedzierski

Contents

Part III: Providing Support for Project Goal Achievement

Introduction

For almost a decade, the Top 500 Project Management Benchmarking Forum has studied project management competencies. The results of the effort are presented as two companion volumes of global standards and best practices. *The Superior Project Manager* identifies, provides research validation for, and weights best practices that are predictors of *project manager competency. The Superior Project Organization* addresses global standards and best practices of *project and host organizations.*

The Superior Project Organization: Global Competency Standards and Best Practices, deals with the practicalities of implementing and operating project organizations. When this effort started in the early 1990s little information was available to assist organizations in developing the project organization. By necessity, much of the information contained in this document evolved as a result of holding benchmark forums three or four times per year, then having participants return to their organizations and implement the theories of the forum, and then meet again to discuss results and experiences. The information learned on the job was validated by conducting a literature search of related empirical research.

Through the experiences of benchmarkers, the phenomenal benefits resulting from implementing professional project management in large functional organizations were documented and publicized throughout the profession. It quickly became apparent that the professional management of projects spanning multiple functional areas resolved or minimized many of the problems inherent in large organizations. Many of the lessons learned from the forums were immediately implemented and have become standard operating practices in best practices project organizations. It is hoped that this volume will introduce other new concepts and result in further improvements.

The Top 500 Project Management Benchmarking Forum was established as a nonprofit organization. One of its missions was to disseminate knowledge for the benefit of industry in general. Its first effort was to publish the book *Best Practices of Project Management Groups in Large Functional Organizations,* which described and detailed the results of early research and outlined thoughts about best practices. The current volumes represent an expansion of that work. Included are global standards and best practices that represent generalized guidelines for the most effective ways to implement and execute project management groups in large functional organizations. It is hoped by the participants in the benchmarking forum that the two volumes of global standards serve as an incentive for organizations to gather and share additional information about best practices and competencies.

0.1. COMPETENCY STANDARDS ASSUMPTIONS

The assumptions under which the forum discussions were held and the research conducted and presented as standards and best practices are as follows.

0.1.1. Large Functional Organizations

The focus is on the use of project management in large functional organizations. For example, the company size of benchmarking participants ranges from about $1 to $155 billion in sales with the mean being about $75 billion. The average number of projects for the project organization is 74, with one project group executing 7500 projects a year. The myriad of smaller matrix-type and informal, single-department—oriented projects are not covered. Nevertheless, most of the concepts described apply equally well to those smaller projects.

0.1.2. Full-Time Project Manager

The primary interest of the forum and this book is on larger projects that require a full-time manager. The assumption is made that the probability of project goal achievement is maximized when projects are executed by qualified project managers who are given authority and accountability to accomplish the project objectives within established parameters.

0.2. RECOMMENDATIONS FOR THE USE OF COMPETENCY STANDARDS

As the study has progressed, participants have developed recommendations regarding the most effective implementation of the standards and best practices. They are to

- Develop a strategic framework on which to direct one's organizational and personal approach to project management. The framework should be based on validated best practices of other organizations as supported by empirical research. The strategic model also provides a tool to resist the temptation to react to organizational pressures and agendas of other stakeholders.
- Insist on rising to the highest level of project management quality and effectiveness.
- Make project management and leadership character, professionalism, and methodology the mantra to guide individual project organization decisions and develop objectives and courses of action.
- Prioritize the best practices in terms of the greatest gain they will generate for the organization.
- Develop a training and development program to align the organization and individual project managers with the concept of best practices.

0.3. FACTORS IMPACTING PROJECT GOAL ACHIEVEMENT

Project goal achievement is influenced by four basic groups of factors: the superior project manager, the project office organization, the host organization, and the external environment. As depicted in Figure 0.1, the impact of these various factors can be weighted. For purposes of generalized discussion, the project manager influences approximately 50% of project success, the project office organization about 20%, the host organization 20%, and the external environment 10%. Note that these percentages can change dramatically depending on the situation. For example, for projects conducted in foreign countries

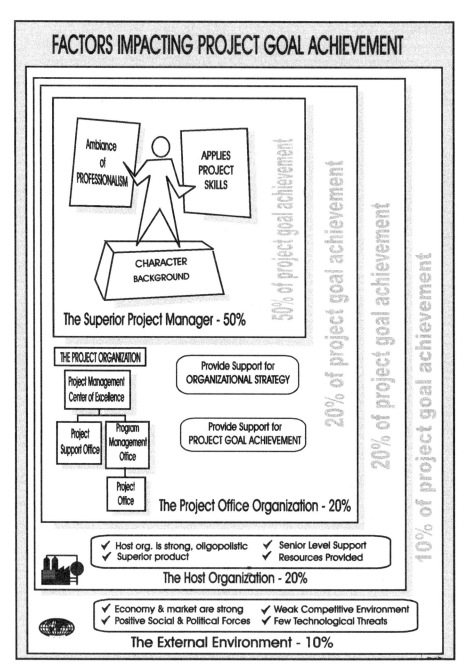

FIGURE 0.1

the external environment could be *the* major influencer of project success.

In most situations, the superior project manager is the single most important influence impacting successful project goal achievement. However, as implied by Figure 0.1, consistent, efficient, and effective project management success necessitates input from, and a balance between, all the key components. Although any one component can be strong and compensate for weakness in other areas, the probability of goal achievement is maximized when project teams have a goodness of fit with the organiztional and environmental components.

The consistently superior project team will ideally be found in a team-friendly organizational environment conducive to goal accomplishment. When all the factors influencing project goal achievement are in balance, everything required for project success is provided. At the same time, there is no excess of resources, complexity in structure, procedures, tools, or resources. As a result, project speed, efficiency, and effectiveness are maximized.

The need for unity between critical performance components often necessitates give-and-take and compromise. In all cases, the project groups are required to make a positive and clearly identifiable contribution to the organization's bottom line and strategy.

Standard: **Project goal achievement probability is maximized when the project manager, project office organization, host organization, and external environment components work in unity to achieve common objectives.**

References: Forrester and Drexler, 1999; Campion et al., 1996; Banner and Gagnle, 1995; Mohrman, Cohen, and Mohrman, 1995; Shonk, 1992; Hackman, 1990.

0.3.1. The Superior Project Manager

> **Standard:** **Stakeholders recognize the superior project manager as the single most important factor impacting project goal achievement.**

The competencies and best practices of the superior project manager are detailed in *The Superior Project Manager: Global Competency Standards and Best Practices.* That volume investigates and reports the experiences and conclusions of benchmarking practitioners combined with historical research that the superior project manager is *the most important component* influencing the probability of project success. This is true in virtually all project environments surveyed, studies researched, and historical literature reviewed. The leader is responsible for the project's entire operations. Team outcomes, strategies, and effectiveness are viewed as reflections of the values and cognitive basis of the team leader. The project manager leads the team in assuming project ownership, responsibility, and accountability.

The superior project manager has the ability to overcome nearly any *controllable* obstacle. The project manager also is the key factor in recognizing and mitigating the impact of *uncontrollable* events. The project office organization might be nonexistent, the host organization could be weak, and adverse conditions might be encountered in the external environment. Nevertheless, the superior project manager will minimize these obstacles and work to achieve project goals.

Numerous researchers conclude that the project leader is the pivotal figure in the determination of project success. The research of Clark and Fujimoto deduces that the superior project manager commands greater respect, attracts better team members, and keeps the group focused and motivated. The project leaders serves as the linchpin between the project team and stakeholders. The superior project leader is the key element in lobbying for resources and working within complex political and cultural environments. They protect the group from outside interference. The superior project manager commands greater respect from stakeholders, attracts better members to the team, and keeps the group focused and motivated.

Krech and Crutchfield observe that by virtue of his special position in the group the leader serves as the primary agent for the determination of group structure, group atmosphere, group goals, group ideology, and group activities. The leader is always the nucleus of a team's ability to perform and the focus of group change, activity, and process.

Although the leader is an important element, a large portion of project success is a reflection of factors outside the control of the project manager. A project that is doomed to failure will still be doomed to failure no matter how competent its leader is. The same research that emphasizes the importance of the project manager can also be interpreted to show that approximately 53 to 65% of the factors impacting project success are outside the sphere of control of the project manager. Examples of factors beyond the control of the project manager are market conditions, competition, government edicts, laws and regulations, and even political events within the organization. Research conclusions also caution that filling the central leadership position within the group is not assurance that the leadership is competent.

References: Forrester and Drexler, 1999; Brown and Eisenhardt, 1995; Clark and Fujimoto, 1991; Hise and McDaniel, 1988; Hambrick and Mason, 1984; Cooper and Kleinschmidt, 1987; Bass, 1982; Krech and Crutchfield, 1948.

0.3.2. The Project Office Organization

As discussed and detailed in this volume, the project office organization integrates the work of the project groups and provides coherency between the project teams, the host organization, and the external environment. It has primary responsibility to ensure that corporate strategy is attained at the tactical or project level. Most project office organizations are also charged with increasing the probability of project success by providing support to the project manager. The project office organization ensures that the project teams are provided tools and resources. Compensation and reward systems, performance measurement and appraisal, training, and mentoring nurture professionalism. The project office organization strives to ensure that the application of organization policies supports project success.

0.3.3. The Host Organization

Project goal achievement is impacted by the host organization (e.g., the company or government agency in which the project group resides). The most consistently successful project teams reside in organizations that are supportive of project teams. Greater project success can be predicted if the project and project organization receive senior-level support, adequate resources are provided, the host organization is financially strong, the market is oligopolistic with few competitors, and the project group is assigned a superior project or service to develop. The host organization provides policies, strategies, and a philosophy of management and behavior to the project organization and teams. The superior host organization structures the project organization for project efficiency and effectiveness.

0.3.4. The External Environment

Many remote forces outside the influence of the project team, the project office organization, and the host organization impact project success. A strong economy and market give support to all the organization's activities, as does a favorable social and political environment. A weak competitive arena combined with few external technological threats has a positive impact on project goal achievement. All these forces shape the external environment and can dramatically improve or seriously hinder the ability of the project team to attain its goals.

The external environment heavily influences projects conducted in foreign environments. Cultural values and mores impact acceptance of team personnel and relationships between stakeholders. The political environment will determine the government's attitude and approach toward the project. Varying approaches to human resource management will affect availability, quality, and relationships with labor. Different financial systems and markets including accounting and taxation regulations affect all elements of the project involving cash flow and funding. The legal system will define the contractual relationships between the parties and attitudes toward personal and property rights. General topography, working environment, and logistics influence the mechanics of executing the project. All these factors could easily result in the external environment representing the dominant element impacting the probability of project success.

0.4. OVERALL CORE SUCCESS FACTORS IN PROJECT MANAGEMENT

A review of the research literature indicates that there are several overall *core factors and competencies* that have the greatest impact on the probability of project success as well as the most influence on project productivity and efficiency. An organization could maximize its probability of consistently attaining project goals by immediately implementing these competencies and developing strategies to optimize the impact of the associated factors. The core factors and competencies relate to the following topics.

0.4.1. Professional Project Manager

The project manager has direct influence over 34 to 47% of project success according to various studies. An organization can maximize its probability of consistently attaining project goals by recruiting, developing, nurturing, and retaining superior project managers.

0.4.2. Multifunctional Teams

Many of the phenomenal increases in revenue and reductions in cost associated with projects are the result of utilizing multifunctional project teams. By definition, the teams consist of representatives from each functional area (e.g., engineering, marketing, production, and finance and accounting) that are impacted by the project. The primary output of multifunctional teams is *increased speed* compared to performing projects in a sequential manner.

0.4.3. Senior-Level Support

The support of senior management translates to project success. Resources are provided and organizational cooperation is increased.

0.4.4. Competitive Product or Service

A key to project success is selection of a competitive product or service. Even the best managed project teams are burdened if they are working on a product or service that is deficient or noncompetitive.

0.4.5. Strong Market Demand

A strong economy and market demand for the project's product or service generates interest and positive acceptance.

0.4.6. Structured Project Approach

Use of analytical, structured, broad, and flexible methodology results in a predictable approach and project outcome. The methodology should be the *least complex* that is appropriate to efficiently and effectively attain the project goals.

0.5. MEASUREMENT OF COMPETENCIES AND BEST PRACTICES

To be accepted as valid, standards and best practices should be observable and measurable. The measurement aspect of the equation

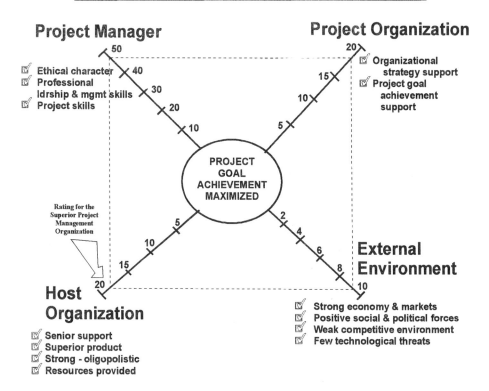

PROJECT GOAL ACHIEVEMENT COMPETENCY WHEEL

Project Manager

- ☑ Ethical character
- ☑ Professional ldrship & mgmt skills
- ☑ Project skills

50 40 30 20 10

Project Organization

- ☑ Organizational strategy support
- ☑ Project goal achievement support

20 15 10 5

PROJECT GOAL ACHIEVEMENT MAXIMIZED

Rating for the Superior Project Management Organization

5 10 15 20

Host Organization

- ☑ Senior support
- ☑ Superior product
- ☑ Strong - oligopolistic
- ☑ Resources provided

2 4 6 8 10

External Environment

- ☑ Strong economy & markets
- ☑ Positive social & political forces
- ☑ Weak competitive environment
- ☑ Few technological threats

FIGURE 0.2

is accomplished by questionnaires and subjective evaluations. The competency wheel is a tool that graphically displays the numerical results and visually depicts the potential for project goal achievement (see Figure 0.2). The specific rankings in Figure 0.2 correspond roughly to the results of historical research that qualifies the importance of each category. Also shown is a dotted "baseline" showing the average ranking for the superior project organizations as defined by results of the Top 500 Project Management Benchmarking Forum questionnaire. The spokes of the competency wheel represent each of the broad areas impacting project success: the project manager, the project organization office, the host organization, and the external environment.

For those desiring additional detail or wishing to rate each of the specific competency components, the master wheel can be decom-

FIGURE 0.3

posed into subwheels. As an example, Figure 0.3 shows the competency subwheel for the project organization. Each of the components of the project organization competency subwheel is covered in this volume. Also included are self-rating questionnaires at the end of each section. The total values of each of the subwheels roll up to equal the overall project success competency wheel.

One can use the competency subwheels to *subjectively* evaluate their organizations in the same manner as for the overall project success wheel. In addition, questionnaires located throughout the text can be completed to determine a detailed *numerical* organizational rating for each element of the wheel. The competency wheel approach is flexible and adaptable to different situations. Users can change the relative values of each competency to reflect a particular project group, company, organization, or global culture and location.

The competency wheels are relatively easy to develop, use, and interpret. They represent a systematic way of determining whether an organization falls within specified boundaries, predetermined benchmarks, or measurement standards. Competency wheels are visual and provide a quantitative and comparative format. They can be used to monitor changes in organizations, to track the results of corrective or training actions, and to compare organizations in dissimilar work environments or time periods. They are suitable for statistical identification of variances as well as subjective evaluations.

References: Moretti et al., 1991; Hornaday and Aboud, 1971; Likert, 1961.

Part I

Overview and Background of the Superior Project Organization

1

Functions, Activities, and Best Practices

1.1. THE PROJECT ORGANIZATION FUNCTIONS AND ACTIVITIES

The project organization (often called the "project office") provides support for two basic groups of activities: it implements the strategy of the host organization as it pertains to projects, and it assists project teams in their efforts to attain project goals through speed, efficiency, and effectiveness.

The project organization supports *organizational strategy* (see Figure 1.1). It maintains focus on project and project portfolio goals as a reflection of the host organization's goals and objectives. The project organization seeks to have the project team involved in projects from inception to termination. It ensures that the portfolio of projects is structured to maximize returns while minimizing portfolio risk. Included in portfolio management is performance evaluation of projects and project managers. As a part of the host organization's strategy, some project organizations are charged with the task of managing individual projects.

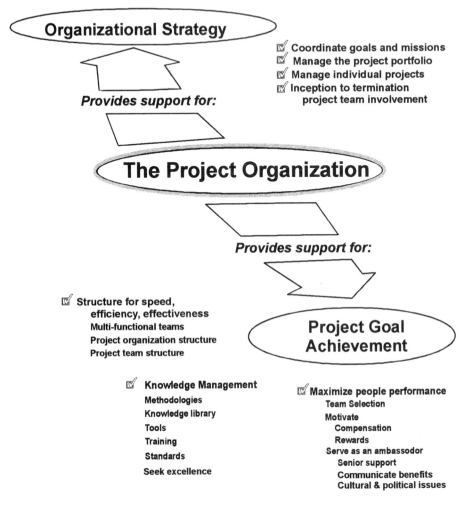

PROJECT ORGANIZATION FUNCTIONS

Organizational Strategy

☑ Coordinate goals and missions
☑ Manage the project portfolio
☑ Manage individual projects
☑ Inception to termination
    project team involvement

Provides support for:

The Project Organization

Provides support for:

☑ Structure for speed,
  efficiency, effectiveness
Multi-functional teams
Project organization structure
Project team structure

**Project Goal
Achievement**

☑ Knowledge Management
Methodologies
Knowledge library
Tools
Training
Standards
Seek excellence

☑ Maximize people performance
Team Selection
Motivate
  Compensation
  Rewards
Serve as an ambassodor
  Senior support
  Communicate benefits
  Cultural & political issues

FIGURE 1.1

    The project organization supports *project goal achievement*. The project organization is structured to encourage project speed, efficiency, and effectiveness. The multifunctional project team approach and its associated management of project activities in parallel rather than sequentially are embraced. Team member duration is managed to ensure fresh thinking while maximizing the benefits of experience.

    The project organization supports project goal achievement through the management of knowledge. Provided are structured and predictable methodologies. A knowledge library serves to transfer knowledge to future projects and ensures that project management is continuously improved. The library typically preserves historical lessons learned, project journals, project charters, and project plans. Also included are project related tools, templates, and methodologies.

The project organization searches for tools and software to aid the project teams. Training focuses on the improvement of competencies, best practices, and skills. Emphasized are on-the-job training, mentoring, and courses that improve character, professionalism, and project specific skills. The project organization seeks excellence through benchmarking with other superior organizations and setting high standards of performance.

People performance is maximized. Team selection receives high priority as does the nurturing and personal development of the project manager and team members. The project organization provides motivational direction for project leadership and team members. Compensation is administered as are performance rewards and recognition. The project organization plays an ambassadorship role. It obtains resources for project teams, ensures senior level support, engages in public relations, assists with sensitive political and cultural issues, and communicates the benefits of projects, the project group, and the overall portfolio to management and other stakeholders.

1.2. THE TOP TEN PROJECT ORGANIZATION BEST PRACTICES

Best practices are observable, measurable, and repeatable actions that maximize the probability of project goal achievement. They are generally agreed upon by industry experts and supported by independent, objective, and empirical research. In the case of the standards detailed in this book they have a measure of authority from the Top 500 Project Management Benchmarking Forum.

Based upon research conducted by the benchmarking forum, supported by historical studies, and validated by industry experience, the top ten best practices of project organizations that impact project goal achievement are listed below.

1.2.1. Provide Senior Level Support

Project office success is dependent upon support from senior levels of management. Senior level support includes having the project organization report to an executive level with authority over multifunctional activities. In cases where projects are strategically important, the project manager may hold a senior level position.

Senior level support is a key element of project organization success for numerous reasons. Often the project organization represents dramatic change for the traditional functional organization. Senior level guidance minimizes cultural problems and encourages acceptance. Public support by senior management gives the project organization and individual projects implied authority. For large organizations or those with global operations, the implementation of global standards, practices, uses of methodology, templates, and tools is eased. Resources are obtained with minimal effort. The probability of project goal achievement is increased because senior management ensures unity of direction by communicating the organization's strategy and vision directly to the project team. Assurance that the project organization efforts are compatible with existing organizational strategy is maximized.

Best practices project organizations develop formal procedures to ensure that senior management support is maintained and nurtured. In particular is the communication of the benefits of the project management organization, project portfolio, and specific strategically important projects.

1.2.2. Superior Project Managers and Team Members Selection

As in the case of putting together a sporting team, selection of the players is critical. A team composed of professionals will have an inherent advantage over a team of amateurs or average players. Some superior project organizations consider the selection of the project manager and associated team members as the most critical activity impacting project success. One company even ranks specific projects as candidates for their project portfolio according to the availability of team members. They assert that people are their most scarce resource rather than money.

The project manager is the key team member to select. This one individual is directly responsible for between 25 and 75% of the probability of goal achievement according to various studies. The impact of the project manager is so dominant that it tends to bias the research. For example, researchers searching for best practices that impact project goal achievement sometimes find that projects are successful even when they are imbedded in inefficient organizations, are in an environment that is not conducive to success, or are in adverse economic conditions. Invariably in those situations, the researchers find a superior project manager. The conclusion is that the superior project manager can compensate for almost any disadvantage or environmental or organizational handicap.

Management often demonstrates an intuitive recognition of the value of the superior project manager and team. When a project is in trouble, one of the first steps sometimes taken to reconcile the situation is to change the project manager and critical team members. In situations where the critical path must be accelerated or the duration of the project compressed, a proven method of fast tracking the project, without increasing costs or reducing scope, is to insert higher quality project manager and team members. Particularly on high tech projects where the technology is changing as the project is unfolding, and the customer wants the latest technology, the project manager is the key element that provides flexibility to adjust and adapt to changing conditions.

1.2.3. Multifunctional Team Structures

Multifunctional teams are composed of team members from all functional areas involved in the project. For example, a new product development group might include design engineers and marketing, production, and finance personnel. Many best practices organizations have recorded outstanding results by including outside suppliers and vendors in the process. Multifunctional team members work simultaneously, in a parallel effort, on each of the functional aspects of the project. Discarded is the traditional sequential approach where each functional area performs its activity and then passes the project to the next functional area.

Research of the phenomenal bottom line improvements resulting from the implementation of project management best practices finds that they are often largely the result of taking a multifunctional approach. Researchers observe that multifunctional teams recognize current and downstream problems and mistakes earlier than other forms of team structure. Representation from more organizational functional areas results in better communications. Conducting projects in parallel rather than sequentially means that the team completes more projects in a given amount of time. As a result, asset and investment turnover are increased and project team cost related to output is minimized.

Multifunctional project management is conducive to fast tracking the project even more by encouraging the overlapping of project phases (e.g., a new project phase is commenced before the preceding one is completed). Speed is increased since each functional team member or group is working at its own pace on a series of activities. Involving multifunctional representatives early in the process reduces wait and slack time between phases and activities. The same benefits accrue from large projects with numerous subprojects where each subproject is working as an independent unit. In all these cases, it is no longer necessary to wait at each phase gate or milestone for all the various project elements to complete the tasks associated with the particular phase.

Benchmark forum attendees advise that a common problem is that organizations *think* they have multifunctional teams when in fact key functional areas are excluded. For example, it is common for multifunctional project groups to include all functional elements except control of the money (accounting and finance) and control of personnel (selection, pay, and promotions). Without these key components the authority of the project manager is hampered as is the probability of successful goal achievement. The situation is similar to informing the CEO of a company that he or she has no authority over the budget nor the authority to hire or fire people. Benchmarkers assert that the organizational leader needs authority over *all* functional areas in order to maximize the potential for project success.

1.2.4. High Standard of Truthfulness

Whether investigating the success of an individual or organization, the single most important factor that impacts goal achievement is truthfulness. On the average, the more truthful the individual or organization, the higher their rate of success. The project organization has considerable impact on building trust throughout the project team. It serves as the role model and establishes the standards for integrity.

Numerous studies verify the importance of truthfulness. One investigation concluded that truthfulness is such an important characteristic that it compensates for major shortcomings in other areas. For example, when asked about the quality of their leader, employees would make such statements as, "My leader has flaws but he or she always tells me the truth. I like that."

The results of trust can be measured in the form of increased sales and reduced costs. In addition, truthfulness lowers risk and generates a myriad of organizational and personal benefits. Project organizations with the reputation for honesty have a natural advantage in the sales arena. The old adage applies, "People like to buy from people they trust." Controlled tests of small groups conclude that project teams composed of members who have a high degree of trust are more efficient and achieve greater output. The trustful groups spend more time on productive activities and less time on paperwork, writing agreements and contracts, and building control mechanisms. Truthfulness increases communications accuracy and openness and makes possible a more free exchange of knowledge and information. It results in faster response time when adjusting to new situations not covered in the original contract or project plan. Speed to market is also increased. Truthfulness lowers project risk and uncertainty by increasing the predictability and behavioral consistency of individuals and teams. Dependable behavior over time and in various environments and cultures gives team members the ability to implement courses of action that are reflective of leadership strategy and philosophy. Trust is self-curing or self-healing through constant monitoring, quick correction of inaccuracies, and repair of trust as tensions arise. Trusting individuals promote interdependence and cooperation. Truthfulness reduces legal fees.

1.2.5. Ambassadorship

Researchers agree that a clear behavior pattern of successful project organizations is emphasis on ambassadorship and its associated activities: impression management, public relations, promotion of and lobbying for project interests and needs, and involvement in political issues impacting projects and the project organization. Benchmark forum participants stress that a driver of project speed and success is the ability of the project team and its supporting organization to successfully interface with an entrenched host organizational culture and bureaucracy.

The success of project management groups in large functional organizations is not simply the result of open armed acceptance by peer groups and superiors. It necessitates an effort on the part of the project organization to communicate the needs, interests, accomplishments, and benefits of the project organization, teams, and portfolio and individual projects to outsiders and stakeholders. Ambassadorial activities enhance project visibility and outsiders' perceptions of team competency. The role includes performing communications and acquiring resources for the project teams. The superior project organization serves as a buffer or "gatekeeper" between project teams and the external political environment.

The ambassador function can have a significant impact on reducing project failure as well as ensuring its success. One researcher concluded that one-third of failed projects are the result of factors associated with cultural conflicts, i.e., politics. Another researcher determined that a major problem encountered by highly technical project managers is interfacing with the host organization and other non-technical functional areas.

1.2.6. Competency Based Training

Research indicates that one of the highest returns related to investment dollars results from training. For example, sales techniques training for sales people can double sales immediately after the instruction. In like manner, emphasis on training is an area that differentiates the superior project organization. There is a strong positive correlation between training and project goal achievement.

Benchmark forum participants agree about the value of training for project managers and team members—as long as it produces results. Forum respondents view training as an investment in better performance and recommend that training should focus on improving project manager best practices and competencies. As a guide, the companion volume *The Superior Project Manager: Global Standards and Best Practices* addresses project manager best practices and competencies and their measurement and application. It also serves as a guide to identify training needs and develop training programs for individuals as well as groups of project managers.

Most companies in the Benchmarking Forum offer five to seven project management specific courses to project managers and project team members. All of the training programs surveyed cover project management methodologies, tools, and techniques of managing projects. All also emphasize leadership and managing people. Some companies that employ predominately technical people feel that leadership and people management training is so important that as many as three courses out of seven convey this type of knowledge. Companies that are globally diverse offer courses in international project management. These courses are considered of particular of value because they cover topics not normally addressed in other project management courses, but which can have strong impact on the project leader's success. Examples of such topics are the global and cultural impacts of religion, politics, race, manners, dress, appearance, mannerisms, speech, and language.

The topic of training includes preparation for and successful completion of the Project Management Institute's PMP (Project Management Professional) certification. The certificate is held in high regard, is globally recognized as reflective of a valid universal approach, and signifies that the holder has a minimum level of experience and knowledge about project management.

Best practices project managers and their companies also emphasize *maintenance* of project management competencies and continuing education. Some participant companies require as much as 40 hours of training each year.

1.2.7. Standardized Methodologies, Tools, and Templates

The development of standardized methodologies and tools is based on the theory that there is a body of universal best practices that apply to every large project no matter the type or location conducted. The objective of *Global Competency Standards and Best Practices* is to address and identify those best practices and competencies. The employment of a standardized set of project specific best practices and methodologies enhances project speed, efficiency, and effectiveness. Standardized methodologies and tools make the management approach to projects predictable, increase stakeholder confidence, and paint an image of professionalism.

As companies become larger and more globally diverse, predictability of project manager performance assumes increasing importance. For large global organizations standardization enables projects to be conducted in any location and cultural or organizational setting, and to deal with any technology service or task, in a predictable manner with minimal amounts of supervision and contact with central organizations.

Structured methodologies and tools increase project speed. The structured approach to managing projects (as compared to a random or ad hoc approach) historically has been found to be the consistently fastest way to achieve goals, particularly on large projects.

Use of a predictable methodology reduces costs. Efficiency is improved by emphasizing a repeatable and predictable approach. It isn't necessary to reinvent the project leadership body of knowledge wheel every time a project is conducted. If the project manager is changed in the middle of the project the new project manager can step in, understand the project management process, and become effective with minimal delay.

Effectiveness is improved because there is a higher probability of goal achievement. The procedural approach ensures that all steps are taken to attain project success. Following a standardized approach improves project management predictability. Stakeholder confidence about project leadership and the manner in which the project will be run is improved. A professional image is presented, particularly when making presentations about project and portfolio options, alternatives, concepts, opportunities, and status. When dealing with technical and highly complex topics, the structured approach tends to encourage simplification through a step-by-step, graphical, modularized, and phased view. Improved understanding; more easily understood presentations; and better initiation, selection, planning, execution, and termination result for the project itself.

Note that there is a danger in specifying the development and application of a structured methodology as a best practice. Some organizations assume that if a little methodology is good, then a lot should be even better. Benchmarkers counsel that an "appropriate" level of methodology should be applied. The judgment of the forum is that the broad variety of projects necessitates flexibility in the management approach and that the superior project manager applies the *simplest methodology necessary to achieve project goals*. Most best practices project organizations have a few core documents such as the charter, project plan, variance reports, and termination checklist. A library is maintained of other optional templates and tools.

1.2.8. Maintenance of a Knowledge Library to Transfer and Expand the Body of Project Knowledge

Best practices project organizations continuously improve project management performance by transferring knowledge from one project to the next. The knowledge library increases project speed, efficiency, and effectiveness. Time to develop the project is reduced, prior problems are avoided, and the new project builds upon historically successful approaches.

Maintaining a library of information gleaned from each successive project attains these benefits. Examples of information retained are the project charter (information used to justify and select the project), the project plan, project journals, and lessons learned. Project managers use the information to select, initiate, plan, execute, and terminate new projects. Where large knowledge libraries exist, the process of planning a project is as simple as finding a prior project that is similar and building the new project on that foundation.

1.2.9. End-to-End Project Involvement

The experience of benchmarking participants concludes that one of the best ways to improve project goal achievement is to involve the project group in the project management process throughout its entire life. Some high technology companies even send the project manager out with sales personnel during the final definition of the contract. In complicated projects, the practice increases the understanding of the customer as well as the sales personnel about what will be delivered and the manner in which the project will be conducted. The involvement of the project group in the initiation and selection process ensures that the project can be performed and the customer has a better understanding of the manner in which the project will proceed.

The research data conclude that early project organization involvement results in definition of the project before large amounts of money are spent and stakeholder positions become entrenched. There is better understanding of customer requirements, technical feasibility, manufacturing, and marketability of the product or service. Trust and commitment of stakeholders is gained early, and false starts are avoided. Response time is reduced because future changes are minimized.

1.2.10. Management of the Organization's Project Portfolio

Best practices project organizations recognize the value of the portfolio approach in managing groups of large, multifunctional projects. The body of knowledge utilized is the same as that applied to manage any group of assets and originates in the world of financial management. The concept of portfolio management can be visualized by picturing the difference between a disorganized sporting team of independent players (i.e., individual projects) compared to a coordinated team (i.e., a portfolio) that is structured to work together and generate synergy in attaining goals.

In portfolio project management, the project organization is charged with optimizing the strategic value of the portfolio of multifunctional projects. The tactic of maximizing overall portfolio financial return (or goal achievement) while minimizing risk attains the objective. Organizational risk can be reduced by strategically diversifying the portfolio of projects. Organizational strategy can be more focused and prioritized as a result of ranking projects according to potential financial return or goal achievement, risk, and strategic aspects. In addition, portfolio management makes possible synergy of effort where one project can assist as well as receive help from others.

Portfolio project management improves the ability of the project organization to measure performance. Before portfolio management, the tendency of organizations was to measure individual project performance independently of other projects. Portfolio management recognizes that whether it is for an individual, a project, or an organization, it is easier to evaluate performance when there are numerous similar examples against which to compare. No matter if the measurement metric is any of a variety of factors, such as strategic importance, profitability, risk, cost, or the project manager's performance and application of best practices, portfolio management provides a measurement metric in the form of a group average or baseline.

When portfolio management is introduced in a functional organization, the first step is usually to take an inventory of all large, multifunctional projects. Benchmarkers universally report that anomalies quickly become visible and more easily resolved. Duplicate projects are eliminated, projects that never terminate are terminated, and projects that make no contribution to the organization are purged. The portfolio approach spotlights projects that are inferior and contribute little to the organization. As a result they can be audited and terminated in a pragmatic fashion.

2

Starting and Promoting
the Project Organization

2.1. IMPLEMENTATION STEPS FOR THE NEW PROJECT ORGANIZATION

Over half the participants in the Benchmarking Forum report that their host organizations have a formal project group or office. The percentage is increasing as success stories accumulate that relate the positive results from applying professional project management in large functional organizations. Many people who join the Top 500 Project Management Benchmarking Forum are investigating project organization related best practices or hoping to implement a project management office or organization. The first information they generally request is a description of the steps taken by other best practices organizations in establishing the project group. In response, the consensus of the benchmarking group is that the following actions are most commonly followed.

1. Define the organizational needs and problems to resolve.
2. Find or create a project office champion or executive sponsor.

3. Build a business case analyzing the benefits of a project orga-
    nization and communicate the results with senior manage-
    ment.
4. Create a center of excellence
    a. Communicate the benefits of project management to
        other organizational areas.
    b. Make project management multifunctional
    c. Train project managers
    d. Develop standardized methodologies, procedures, and
        templates

2.1.1. Define the Organizational Needs and Problems to Be Resolved

Before implementing the project office concept, Benchmarking Forum
participants emphasize that there should be a clear need and measurable potential benefits. Specifically, the project organization should
represent prospective improvements in the organization's bottom line
(i.e., profit) through increased sales and reduced costs, a better ability
to attain project goals, and/or improvement in implementation of the
organization's strategy.

Benchmarkers report that recognition of the need for a project
organization generally starts with the functional organizational
structure and its associated problems. Management recognizes that
large projects and activities that span functional areas aren't working
as well as they could. Frustration levels may be high because projects
often seem to be in trouble. They fall short in delivering the benefits
promised or are late, are over cost, and deviate dramatically from the
original project plan and promised functional and technical specifications. Many projects never seem to end or terminate and have an
unquenchable appetite for scarce financial and human resources.

Sometimes external events signal the need for improvement.
As the pace of technological change accelerates, pressure builds for
faster project performance and shorter lead times to market. Even
as corporations react by intensifying their efforts, competitors may
consistently be quicker in getting products to market and responding
to market opportunities.

In other cases, projects subject to high public and stakeholder
scrutiny might be deemed failures by the press. Stock values and
company profits may decline as could the organization's image of expertise and professionalism. Often such projects represent major
commitments of funds and long-term strategic dedication.

The results of researcher Tom Ingram's investigations support the experiences of the benchmarkers. Ingram evaluated 60 project failures that were significant enough to be reported in the popular press. The companies studied were recognized globally as experts and masters of the functional activities and technologies for which their companies were known. In most every case, it was project management with which they were unfamiliar. Ingram concluded that when functional companies face major problems and losses, it is usually the result of failed projects. Conducting their day-to-day business rarely results in major surprises.

Ingram correlated information about project failures reported in newspaper and magazine articles with stock price declines of the host organization. For the companies he defined as experiencing "major" project management problems, the stock price declined from 20 to 91%. Seven firms analyzed lost an average of 59% of their stock value during the time the projects were underway and receiving adverse publicity. Ingram estimated that had the projects not failed in 13 of the companies evaluated, average net earnings would have increased approximately 6%.

Ingram's analysis of the reported project failures showed that they often occur in organizations that unquestionably exhibit superior professionalism, performance, and predictability of leadership and management practices when executing their functional duties. In these organizations, constant reiterations and improvements in functional activities over decades of time have resulted in high levels of expertise and professionalism. Planning and budgeting for functional activities is precise. In the cases evaluated, these same organizations were often ill prepared to manage large projects requiring multifunctional coordination. Projects experienced failure or veered from planned goals. Problems were accentuated when the projects involved volatile information systems, software, and high technology projects where the technology was changing as the project was underway and the customer wanted the latest technology.

As a result of the problems and need for improvement associated with managing multifunctional projects in large functional organizations, Benchmarking Forum participants placed the project process under a magnifying glass. They identified a host of dilemmas resulting primarily from the organization's functionally based structure as well as its large size.

- **Autonomy.** As organizations become larger and larger, the functional units become increasingly autonomous and independent. There tend to be fewer and fewer opportunities for cooperation and integration of processes at the lower working levels. Functional managers may be reluctant to share projects with other areas or relinquish control when project related activities leave their area.

- **Communications.** Communications between functional areas can become strained as each increment of organizational growth results in the insertion of additional layers of management filters and information barriers. The formal communications track becomes longer and more difficult to traverse.

- **Difficulty seeing the organizational strategy.** As organizational size increases, leadership and its associated vision becomes progressively distanced from the working groups. The addition of each layer of management makes it more difficult for the leader of the company to communicate the organizational vision and goals of the organization to lower levels of management and workers within functional areas. Interfunctional strategic decisionmaking relationships are primarily made at the top—and the top is far away.

- **Inadequate authority.** Most projects conducted within functional departments do not have executive based authority. As a result they have limited access to resources and face numerous challenges when working with other functional and geographic groups.

- **Matrix project organizations.** Functionally based organizations without a project organization tend to conduct projects using a matrix style project team structure. For example, a functional area will appoint a project manager who is tasked to borrow team members and resources from other participating areas. Project team members report to two bosses, the project manager and their functional managers. The constraints of borrowed resources and two bosses hinder the ability to attain project goals.

- **Projects detract from one's full time job**. Most employees in functionally structured organizations already have a full time job to perform. When a project is initiated, the time required to participate is often committed at the sacrifice of one's day-to-day job accomplishment. If individual or group performance is evaluated based on measurable output, such as encountered in sales and production departments, projects are often viewed as counterproductive and a distraction from maximizing the core activity. Where commissions or performance based bonuses are involved, contributing time to the project may even reduce the participant's income. Benchmarkers report that the outcome in such situations is that most projects simply don't get done or are performed in a hurried manner.

- **Unpredictable management approach.** In functional organizations that have yet to fully commit to project manager professionalism, the typical approach by untrained project managers is to execute projects in an ad hoc or unpredictable manner. Project time to completion is excessive. Projects are often conducted in a sequential rather than multifunctional manner. Performance is difficult to evaluate as there are few measurement metrics.

2.1.1.1. Impact of Historical Failures

The frustration of executives in many functional organizations is heightened because they have already tried other approaches to resolving the problems. Often they have reengineered and conducted process improvement programs. In many cases, the efforts resulted in minimal positive results because they failed to resolve the fundamental problems inherent in large functional organizations. One benchmarker commented that the results are like installing a wing on a turtle. Not much is gained because the turtle is already going as fast as it can and will never reach a high enough speed to fly under any condition.

2.1.1.2. A Project Organization Is Not Always Best

Members of the benchmarking forum also emphasize that implementing a project organization is *not always the best course of action.* It is timely to remember that starting a project organization adds cost and results in the inclusion of an additional step and filter in the communications chain. Unless the potential benefits from the project office are measurably greater than its costs, *not* having a project group is a realistic option to consider. Benchmarkers would also testify that it isn't always necessary to have a project office organization to manage projects in a superior fashion. Many organizations without a project organization attain high project goal achievement success rates by nurturing and placing emphasis upon the utilization of superior project managers who apply best practices.

2.1.2. Find or Create a Champion or Executive Sponsor

To ensure success, an executive level sponsor best serves the project organization. At least five research studies agree that senior level support is one of the most important ingredients needed to foster the probability of project success. To strengthen support for the project organization concept, benchmarkers advise building partnerships with executives who impact the group's effort. These individuals can be nurtured and upon witnessing the benefits of project management, then become emissaries of the philosophy. By building support with an executive or executives, the project organization is given direct support and sponsorship as well as inferred authority.

> **Standard: Best practices project organizations build partnerships with and gain support from senior executives.**

Benchmarking participants also say that *without* senior level support it is difficult to successfully implement multi functional project management. Some benchmarking organizations have implemented project groups at lower levels at the initiative of middle management. Usually these were found within a single functional area such as engineering. In many such cases, the project groups found it challenging to move beyond success in their host functional area. Rarely could such a group become truly multifunctional or incorporate people from other functional areas until receiving the support of senior management.

Obtaining senior level support is not an easy task. A survey of benchmarking forum participants disclosed that only 22% said they receive "full" support from the senior officer in the company. Little or no support was recorded by 26% of respondents, with moderate support listed by 52% of the benchmarkers. Nearly three-fourths of survey takers said that top management's commitment is *not* visible to all employees. Approximately 60% indicated that senior management does *not* work to develop the project organization concept.

On a broader scale, benchmarkers relate that project management groups often represent a dramatic change in conduct for many old-line functional organizations. Sometimes the project group is viewed as a threat and embraced with less than enthusiasm. Benchmark attendees relate examples where the project organization achieved increasingly larger and more visible successes, only to be met with amplified resistance and opposition from the traditional organizational culture. Usually the resistance was at upper middle management and vice-presidential levels. In at least two cases, *the project organization was terminated* after receiving publicly recognized successes. Benchmarkers assert that high level executive support ensures that the implementation and acceptance of professional project management is encouraged and even enforced, across the organization.

2.1.3. Build a Business Case for a Project Organization

In nearly every organizational situation reported by benchmarking participants, it was necessary to make a formal proposal to senior management before support could be obtained and the project organization implemented. Discussed in the proposal or presentation were the benefits the project group could be expected to contribute to the organization, the manner in which existing problems could be resolved, and the methods used to satisfy identified needs.

A typical presentation or business proposal details the potential advantages and synergies resulting from combining the expertise of traditional functional organizations with the speed, efficiency, and effectiveness of professionally managed projects. The case is developed that implementation of professional project management in large functional organizations can increase revenue, reduce costs, enhance the performance of individual project managers, improve the image of professionalism in the eyes of customers and other important stakeholders, and provide numerous generalized organizational improvements. The specific benefits usually covered are as follows.

Standard: **Best practices project organizations develop formal presentation that outline the benefits of the project organization.**

2.1.3.1. Bottom Line Financial Benefits

Benchmarkers agree that a major reason senior management is favorably inclined toward professional project management is that it generates *incremental profit for the organization.* Ingram's research infers that profits could increase an average of 6% in companies with superior project management. Dozens of other researchers support his findings with specific examples. A study performed by the *Best Practices Report* concluded that 68% of companies said that increased productivity was the major benefit resulting from project management.

Professional project management generates bottom line improvements by reducing lead time to market and then harvesting the accompanying increased sales, freeing of funds for other investments, execution of more projects with the same resources, quicker response to market opportunities, faster market positioning, and better pricing flexibility. Project organizations also increase profits through attaining lower costs, improving the ability to measure performance, and the quicker identification of failing and marginal projects.

2.1.3.1.1. Reduced Lead Time to Market. Speed based multifunctional project management concentrates on executing and delivering projects in the shortest amount of time. Benefits of the approach are incremental sales gains, an earlier establishment of market position, and improved profit margins. Zangwill's extensive study of "concurrent engineering" or multifunctional project management disclosed numerous specific examples of improvements. AT&T reduced total process time on new products to 46% of an original baseline. At Boeing, parts and materials lead time was lowered by 30% and design analysis time by 90%. Deere & Company cut development time by 60% on the project evaluated. In like manner, Hewlett-Packard cut development cycle time by 35%, and IBM experienced a 40% reduction in design cycle time. In the benchmarking group, a large telecommunications company shortened the average time to complete projects from 52 to 18 months.

- **Incremental sales.** The value of reducing time to market by one day is phenomenal. Each day that time to market is reduced adds one more day of sales. In the case of the telecommunications company that shortened time to market from 52 to 18 months, sales occurred nearly *three years earlier* than the prior approach. The project group estimated incremental sales resulting from the 34 month reduction amounted to $4 billion.
- **Frees funds for other investments**. By generating sales sooner, funds and resources are made available for other investments. The time value of money is maximized. In the preceding case, the project organization calculated that at a 10% interest rate and on a compounded basis, the incremental interest revenue made available to the company amounted to approximately $210 million per year.

- **Execute more projects with the same resources**. Increased project speed means the project team will execute more projects in any given amount of time. Where previously the project team might manage a project for a year or so, now each project is completed in a few months and the team then moves on to another. As an example, the Avraham Y. Goldratt Institute posted a case study on their web site describing Critical Chain Project Management at Lucent Technologies. Not only did Lucent reduce cycle time to produce new products by 50%, but the number of *projects completed more than tripled,* from five projects in 1998 to 16 projects in 1999, with no increase in staff.

  An associated benefit of executing more projects with the same resources is that cash invested in each project is tied up for a shorter period of time. The financial effect is similar to that obtained from increasing inventory turnover. High inventory turnover means that the investment in inventory is small compared to the amount of resultant sales. In the case of project management, resource usage is maximized by (a) making project managers and team members available for other high potential projects, and (b) freeing the financial resources that would have been allocated had the project run for the original extended period of time.

- **Respond to market opportunities sooner.** Benchmarkers whose organizations operate in high technology industries with rapidly changing technologies and market demands report that a key advantage of speed based project management is that it gives the company the capability to react faster. The organization has increased confidence that market opportunities and alternatives can be identified and a response developed faster than competition. The result is that the organization can be consistently first to satisfy market needs and demand.

- **Quicker market position.** Speed based project management and the resultant shortened time to introduction means that market share is obtained early. Competition is less intense and initial demand is higher. A more dominant niche and stronger position can be established.

- **Greater pricing flexibility**. Speed based project groups with a history of being first to market report that the initial high demand and scarcity of the product provide an opportunity for pricing flexibility. Specifically, the new product or service can capitalize on this set of circumstances and maximize profits.

2.1.3.1.2. Reduced Costs. The general consensus of benchmarking participants is that most companies can cut costs approximately 10% by implementing professional project management. Not only do lower costs result in more profit but they also give greater pricing flexibility and competitiveness when contending for business.

An advantage of professional project management is that savings are usually easy to measure and observe. Zangwill's research identified numerous examples of cost reductions resulting from the multifunctional approach to projects. The Texaco Convert Louisiana Refinery was estimated to cost $1.1 billion and require 6 years to design and construct. Zangwill reported that the refinery was finished 1 year sooner and $200 million under budget. At McDonnell Douglas one reactor and missile project recorded 60 specific cost savings from the bid. Boeing Ballistic Systems reduced labor rates by $28 per hour, cost savings by 30%, and material shortages from 12% to zero. Deere & Company lowered construction equipment development costs by 30%. Northrop recorded 30% savings on the bid of a major project.

2.1.3.1.3. Improved Ability to Measure Output. Professional project management lends itself to the establishment of measurement metrics and performance evaluation. Benchmarkers have identified close to 20 different ways of measuring individual project performance. When evaluated as a part of a coordinated portfolio, individual project and project manager performance can be compared to any number of group averages and ranges. Benchmarkers report that until their organizations embraced professional project management, rarely if ever were the benefits of projects or their overall performance measured and communicated.

2.1.3.1.4. Identification of Failing and Marginal Projects Sooner. Professional project management organizations increase organizational efficiency by auditing and monitoring projects. The process ensures that projects remain on track and that ineffective projects are terminated.

Benchmarkers avow that terminating projects is a difficult task in organizations without professional project management. Over 20 companies in the benchmarking forum indicated they *never killed information systems projects* prior to implementing a structured auditing program. Ingram's research found that in organizations without a project management organization that stakeholders were reluctant to step forward and declare the project a failure or in trouble. They were even hesitant to interfere with the day-to-day management of the project. Consequently, projects sometimes continued down the path of failure long after everyone was aware that they were a terminal issue. For example, the project management group at a major telecommunications company reported that an average of 18 months passed before a clearly failing project was even acknowledged by management. After implementation of auditing and review procedures, the project organization identified failing projects *and terminated them* in an average of 6 months.

Implementation of a structured project evaluation and auditing program can reap other immediate benefits for the organization. A major insurance company eliminated 5% of their portfolio of projects that were contributing *nothing* to the attainment of corporate strategy nor had a termination in sight. The scarce financial and personnel resources associated with the terminated projects were made available for other higher return projects.

2.1.3.2. Organizational Benefits

In addition to the bottom line benefits resulting from the implementation of a project organization, there are numerous organizational improvements. The probability of successfully achieving the project's planned goals is maximized, risk is minimized, the attainment of organizational goals receives better support, and project portfolio management maximizes the value of projects as a coordinated group rather than as individual projects. End-to-end involvement of the project organization gives greater assurance of success, scarce entrepreneurial leadership skills are developed and retained, and the multifunctional project approach improves the effectiveness of interdepartmental activities. Expertise and administrative activities improve efficiency by being centralized, standardized methodologies add predictability, there is a single point of accountability which results in faster problem identification and resolution, quality is improved, and there are fewer scope changes.

2.1.3.2.1. Maximizing the Probability of Project Goal Achievement. The project organization ensures the application of industry and research supported best practices and competencies that maximize the probability of project goal achievement. The result is a predictable and professional approach to project management and a higher level of confidence in its successful completion. The *Best Practices Report* survey of the value of project management found that 50% of respondents had experienced significant improvement in completing projects within budget and on schedule.

Implementation of professional project management ensures that projects are conducted with speed, yet are efficient and effective. The effectiveness portion of the equation is particularly important because the project *must be completed and attain its goals*. It is similar to taking an airplane flight from Phoenix, Arizona, to Kennedy Airport in New York City. The aircraft may be the fastest available (speedy), could take the most direct route (efficient), but if it lands at any other airport than Kennedy (the goal), it is not effective. To be effective the project goals must be achieved and the project product or service delivered to the customer.

2.1.3.2.2. Minimizing Risk. Professional project management seeks to maximize financial return *while minimizing risk*. Participants in the benchmarking studies testify that risk associated with projects is inherently greater than encountered in the day-to-day activities of the functional organization. In the beginning of the project there is the risk associated with selecting one project at the exclusion of other potentially good projects. Usually the portfolio selection exercise involves larges amounts of money and resources, the effects of the selection decision remain for a long time, and the results are damaging to the organization and people's careers when wrong. Once the project is approved there are a myriad of additional risks. The team has not worked together in the same environment, the project activity is unique and benefits less from prior learning experiences, deadlines create constant pressure, and there is a much higher degree of personal and team accountability.

Experts in the investment field counsel that risk is a particular problem for individuals and teams new to a discipline. They say that the tendency is to emphasize the amount of *financial return* while minimizing the impact of risk. As experience accumulates, emphasis shifts toward *analyzing the risks* associated with the level of potential return. Professional project management accelerates the learning process with a structured and methodological approach to portfolio and individual project risk evaluation and decisionmaking. The approach ensures that risks are identified, are quantified in terms of probability of occurrence and magnitude of consequence, and a response plan is developed. As a result, there are fewer surprises, financial reversals, and public embarrassments.

2.1.3.2.3. Support of Organizational Goal Achievement. The consensus of researchers and practitioners is that organizations, groups, and individuals that conduct strategic planning are consistently more successful at achieving goals than those who don't. As a reflection of this premise, a key purpose of project management in large functional organizations is to support the host organization's goal achievement strategy. In this respect, professional project management has proven a phenomenal tool. Implementation of a project management group is a key factor that significantly contributes to corporate profit and/or other strategic goals. The structured methodologies ensure that the project team is cognizant of how the project goals, plans, and deliverables interface with and support the organizational vision and strategy. Project organization emphasis is placed upon the effective communication of organizational leadership and goals and the enhancement of goal focus by project teams.

- **Communication of Organizational Leadership Vision and Goals.** By having a project organization that spans functional areas and reports to an executive level, the host organization's vision and goals become closer and easier to communicate to the working groups. It is an important consideration because seasoned project and program managers say the project team *must be* cognizant of how their project fits with the organizational goals, vision, and strategies. When the goals and strategies are not clear, there is a tendency for the group to be unsure of their target and to flounder and lack direction. Improved communications reduce the magnitude of this problem.

  Having a clear understanding of the organizational vision provides the project organization and teams a unified view of the project direction and objectives. It makes decisionmaking faster and more efficient. When team members know the overall vision, there is less need to ask for specific instructions each time new situations, alternatives, opportunities, and dilemmas are encountered.

- **Enhancement of Goal Focus of All Team Members.** Professional project management emphasizes setting clear, measurable and observable goals—and then maintaining focus on the achievement of those goals. Several studies conclude that maintaining a constant focus on the goal has a high correlation with goal achievement. As a result, projects are less subject to scope changes and diversions. They are completed faster and with fewer modifications. The disciplined vision focus keeps the natural disorder of the project under control. Benchmarkers relate that multifunctional projects in large organizations sometimes find it difficult to maintain this focus. When there is no project organization there is a tendency for each functional area to focus on the goals specific to that area rather than those of the entire organization.

2.1.3.2.4. Optimizing the Portfolio of Projects. Approaching the numerous large multifunctional projects in an organization as an integrated portfolio rather than a group of independent projects generates additional revenue, reduces overall organizational risk, and focuses all project efforts on attaining organizational strategic objectives. The project organization is often judged to be the best positioned group to optimize returns from the performance of the portfolio management task.

An overall benefit of portfolio management is that *projects can be ranked* according to potential financial return, risk, and subjective or strategic considerations. An associated benefit is that individual projects can be *compared to a baseline* or group average of the entire portfolio. The ranking process ensures that management is clearly aware of the best as well as the poorest performers. Benchmarkers report that the process minimizes the problem of less important projects receiving inordinate amounts of attention because of politics or nonperformance related variables. More importantly, it ensures that high ranking projects receive appropriate levels of resources and management support. The ability to compare single projects to a baseline, group average, or beta of the portfolio provides a broader range of measuring project and project team performance as well as their relationship to other projects.

A key element in project portfolio management is the selection process. At least three major studies have found a clear positive relationship between proficiency in analyzing and selecting projects and ultimate project success. Best practice selection teams ensure that there is a good fit between the needs of the project and the host organization's skills, resources, and areas of expertise.

An important part of the selection process is the maximization of overall portfolio financial return. A project represents a single financial investment opportunity among a coordinated portfolio of investment activities. Rarely is there enough money and human resources to accept every project that would be beneficial for the organization. Some form of a selection process must be implemented. In this sense, project portfolio management is similar to the management of any other group of sizable organizational assets. Organizational financial return is increased because a pragmatic decision process ensures that each project serves to optimize the overall portfolio potential for financial returns.

Portfolio management offers the opportunity to mitigate risk through pragmatic analysis and studied diversification. Projects can be selected to minimize the effects of external risk such as economic downturn, technological change, and competitive actions and reactions. Risks inherent to the project itself can also be evaluated related to other projects in the portfolio

Portfolio management represents the tactical portion of the organizational strategic plan. The project portfolio can be viewed as a broad based group of assets that can be positioned to achieve maximum overall strategic effectiveness for the organization. In addition, the management of specific projects is more focused because the project team is aware of the relative importance of each project and how it helps achieve the company's strategic objectives.

After project selection, the administration of the portfolio ensures maximal speed, efficiency, and effectiveness. Portfolio project management emphasizes the optimization of the entire inventory of projects. Project organizations execute the portfolio tactical process by taking, updating, and maintaining inventories of existing projects and constantly improving the existing project portfolio through auditing and the subsequent nurturing of healthy projects or pruning of duplicates and nonproductive, ineffective, and failing projects.

2.1.3.2.5. End-to-End Project Involvement Increases Success. Benchmarkers assert that the project organization should manage the project from inception to termination. Specifically, the earlier the involvement of the project group, the higher the probability of project success. Empirical research supports the practitioners' conclusion. In the Gupta and Wilemon study, 42% of respondents stated that early involvement of the multifunctional team is a key element in attaining project success. In most best practices project organizations, the project organization is involved in project selection and project portfolio management. Some organizations report that the project manager even accompanies the sales person when finalizing details on large technical and complex information systems and software projects.

Benchmarkers' experience plus hard research data conclude that integration and involvement of the project group at an early stage in the process results in greater shared commitment, clarified stakeholder needs, and better definition of product specifications before large amounts of money are spent and stakeholder positions become entrenched. Response time is reduced because future changes are minimized. Cooperation of the functional groups and stakeholders is

improved as a result of better understanding of customer requirements, technical feasibility, manufacturing, and marketability of the
product. Trust and commitment of stakeholders is gained early on,
and false starts are avoided. The end result is increased confidence
that the project will perform as planned and a better understanding
of the manner in which the project will proceed.

**2.1.3.2.6. Developing Leadership and Entrepreneurial
Skills.** The project organization serves as an incubator to nurture
and retain aspiring managers who possess the scarce entrepreneurial
seeds that grow into future executives and organizational leaders. A
growing body of research supports the conclusion that the best practices and skills needed to lead and manage an entrepreneurial company are essentially the same as those required to lead and manage
the multifunctional project. Organizational representatives in the
benchmarking forum acknowledge the similarities and comment that
their organizations seek project managers with entrepreneurial characteristics as potential future executives. Some have also voiced the
opinion that the entrepreneur is the type person that formerly was
inclined to leave large organizations and start businesses of their
own. Project management serves as a means to motivate and retain
the entrepreneurial resources.

These views are buttressed by solid research conclusions. Competency research indicates that the characteristics of the superior
project manager are identical to those of the superior *entrepreneurial*
chief executive officer. McClelland's global research defines an entrepreneur as the person who organizes the firm or organizational unit
and increases its productive capability by taking risks. McClelland
proposes that self-confidence and faith in a positive future are key
components of entrepreneurial leadership. The entrepreneurial propensity to take risks has a positive impact on organizational goal
achievement. Self-confidence and faith in a positive future outcome
induce the leader to engage in ventures that less confident people
would hesitate to embrace. The entrepreneurial project leader is more
willing to accept accountability for project success and its associated
rewards and/or potential for failure. The chief executive officer aspect
of the entrepreneur definition infers that the individual is responsible
for all functional aspects of the enterprise or project.

The project environment is conducive to the retention of entrepreneurial resources within the organization. Entrepreneurial individuals can make worthwhile contributions to the organization as
well as exercise their unique talents. A future source of executive
leadership material is assured. The project organization can serve as

the training ground to prepare people for management and executive roles in the organization. Rewards and motivation for risk taking and assumption of bottom line responsibility are provided in the form of remuneration, broader career path opportunities, job satisfaction, and self-fulfillment.

2.1.3.2.7. Improves Coordination of Intergroup Activities.

Implementation of a project organization lends itself to the formation of multifunctional teams. Multifunctional teams work best where there is an independent and high level project organization. Where projects are conducted within functional areas, the various organizational units are sometimes reluctant to turn over authority and resources. The compatibility of multifunctional teams with the project organization is an advantage that, more than any other, results in many of the phenomenal increases in revenue and reductions in cost often attributed to professional project management.

Multifunctional project teams consist of representatives from each functional area (i.e., engineering, marketing, production, finance and accounting) that are impacted by the project. Multifunctional team members work simultaneously, in a parallel effort, on each of the functional aspects of the project. Discarded is the traditional sequential approach where each functional area performs its activity and then passes the project to the next functional area. The primary output of multifunctional teams is *increased speed* compared to performing projects in a sequential manner.

At least three studies conclude that the multifunctional project team is *the* critical element in the rapid attainment of project goals and improvement of project speed. Having a representative on the project team from each functional area improves the coordination of project activities as they flow across functional boundaries. Smoother execution of all phases of the development process is the result. Team members develop empathy for the concerns of other functional areas. There is a deeper understanding about how each of the functional areas is integrated into an effective team to utilize synergy in more effectively attaining goals. Downstream problems are identified and corrected earlier in the project process. As a result, projects are executed faster and more efficiently.

Phenomenal results have resulted from the multifunctional approach. Cross-functional teams have cut development time by 20 to 70%, boosted white collar productivity 20 to 100% and increased sales 5 to 50 times. One research team reported an almost perfect positive correlation of .87 between the use of multifunctional teams and timely project goal achievement.

2.1.3.2.8. Centralizes Expertise and Consolidates Activities. The project organization concentrates project management expertise and combines activities common to multiple projects at one central location. Efficiency is improved and the cost of redundant and duplicate functions is reduced.

A positive result of centralized project organizations is the optimization of management and administration of the project manager resource pool. Skills inventories can be taken that spotlight strong as well as weaker areas needing specific competency based training. Hiring practices can reflect these needs. The end product is a group of project managers and other subject experts that offer a higher level of competence than would be found without the centralized approach.

Most best practices project organizations develop and maintain a knowledge library of information from prior projects including charters, plans, lessons learned, and journals. Using prior project knowledge as a model reduces time required to develop new projects. Diminished is the tendency to repeat errors and the need to constantly reinvent the wheel. The knowledge library also normally includes a standardized methodology along with templates and tools to ensure a predictable approach to project management.

2.1.3.2.9. Results in a Predictable Management Approach. A key output of best practice project organizations is the implementation of a standardized project management methodology. The methodology includes specific best practices, procedures and templates that are appropriate for the broad variety of large projects. The methodologies represent a structured and predictable approach, yet are broadbased and flexible. The most simple methodology appropriate to the project is applied.

The employment of a standardized set of project specific best practices and methodology makes the management approach to projects more predictable. It provides certainty that the large variety of projects in different organizational settings, cultures, and geographical locations will be approached in a similar manner. Cordero's research concludes that exercising a predictable approach to project management tends to simplify team effort and increase project speed. The project team is more efficient. Team members don't need to develop a new methodology every time they lead and manage a project. The best practices, standards, and documentation for each project phase are easier to understand and implement. They provide a common terminology that results in faster and more accurate communication. If the manager of the project is replaced in mid-project, the new project manager understands the process and can quickly adapt to the project specifics. Inexperienced project managers became proficient faster than if they learn the mechanics of team leadership in a random manner.

The methodological approach to project management is evidenced by other benefits. Project effectiveness as measured by the probability of goal achievement is maximized. The procedural approach ensures that all steps are taken to attain project success. In cases where projects are marginal or fail to live up to their initial expectations, the methodology dictates that they will be promptly evaluated and either fixed or terminated. The common frame of reference ensures that projects and project management performance can be evaluated, compared and contrasted between dissimilar projects in various geographical, technical, and organizational environments.

Implementation of a predictable methodology gives stakeholders a measure of confidence about project leadership and the manner in which the project will be run. It presents a professional image. It lends itself to structured presentations and proposals. It aids managers in easily presenting the concepts and status of the organization to stakeholders.

2.1.3.2.10. Provides a Single Point of Accountability. Professional project management is based on the utilization of a project organization and project managers that are responsible and accountable for project performance and goal achievement. With traditional project management the project phases are approached sequentially and span several functional areas. In such environments, it is often difficult to determine who has overall responsibility. Accountability increases efficiency by pinpointing the critical elements of performance. Root problems are identified faster and solutions implemented sooner.

2.1.3.2.11. Improves Quality. When professional project management is implemented the quality of the service or product is typically higher. Usually the improvement comes at no sacrifice of project speed, cost, or effectiveness. According to the research, improved quality is a side-benefit of the increased levels of competency associated with professional project management. Zangwill's investigations identified numerous examples of quality improvements associated with multifunctional project management. At McDonnell Douglas weld defects per unit decreased by 70%, scrap was reduced by 58%, rework cost by 29%, and nonconformances by 38%. Boeing Ballistic Systems decreased the floor inspection ratio by 66% and ran a 99% defect free operation. Northrup reduced defects by 35%, and at AT&T defects declined 30%. Quality improved at Deere and Company to the point that the inspector staff was reduced by 66%. Hewlett-Packard's product field failure rate went down by 60% and scrap and rework by 75%. AT&T's cost of repair for new circuit pack production was cut by 40%.

2.1.3.2.12. Decreases Scope Changes. Benchmarking forum participants report that a constant problem with conventional project management is poorly managed scope changes. Researchers agree. Ingram's investigation of 60 failed projects determined that 70% came in materially late, over budget, *or failed to meet the client's expectations*. The direct and associated costs of the missed targets *averaged eight to ten times the original project budget.* Research of projects investigated by the Standish Group found that *almost all* projects in the study incurred significant scope changes.

Some projects have so many changes that the very nature of the project is distorted. The change process is often continuous with the result that projects become increasingly nebulous and harder to define. Scope changes make it more difficult to measure performance, monitor progress related to the original project plan, and compare current status with prior efforts and milestones. Scope changes generate alterations to the budgeted cost of work scheduled, the day-to-day schedule, the technical and functional specifications, deliverables, and project goals. Accountability is tougher to pinpoint and enforce. Projects become skewed, go off track, and continue in perpetuity.

One reason the project organization is a necessary element in the scope management process is that often the project manager allows and even encourages scope creep and project changes. Benchmarkers say that in almost all cases the project manager is under constant pressure to change the project. The project manager may be susceptible to change requests as a result of making an effort to be cooperative and keep the customer happy. In other cases, the scope changes may be self-serving in nature. Ingram's research identified numerous examples where project managers used uncontrolled scope changes to reduce accountability, disguise performance problems, or continue projects indefinitely when they should have been terminated.

In an ideal world of maximized speed, efficiency, and effectiveness, the scope and technology of the project would be frozen and never changed. In real life, scope changes are a necessary reality. In many organizations scope changes represent opportunities to generate incremental revenue and profit, add value, adjust the project to changing conditions, incorporate improvements, and correct oversights.

Even so, best practices project organizations recognize that the fewer scope changes, the higher the project speed and degree of efficiency and effectiveness. Consequently the superior project organization formalizes the scope change process and manages scope changes rigorously. The results are measurable. Zangwill's research identified several examples of reduced scope change resulting from professional project management. McDonnell Douglas recorded 68% fewer changes on a reactor. Northrup reduced engineering changes by 45%.

2.1.3.3.   Individual Benefits

Individuals are considered valuable assets in most organizations. The project management group can be an instrumental factor in improving the value of those assets. The "Value of Project Management Survey" conducted by the *Best Practices Report* found that 37% of respondents said that improved employee satisfaction was one of the major benefits of professional project management.

The project organization lends itself to the implementation of best practices that reflect observable and measurable standards of performance. Subsequent selection of the best performing individuals for promotion ensures that performance proven individuals rise to higher levels of responsibility within the organization. The skills of prospective project managers can be tested on progressively larger projects. Focused competency based classroom training and increasing responsibilities through the management of larger and larger projects increases the asset value of project managers.

2.1.3.3.1.   Clearly Defines Best Practices and Competencies.   Project manager performance is based upon the application of specific best practices and competencies. All of these are clearly defined, observable, and measurable. As a result it is easier to measure and compare performance of project managers. In situations where the project outcome is unclear or the project manager takes over the project in trouble, the ability to measure competence is much broader and more realistic. Use of best practices as measurement metrics lets the performance of project managers be evaluated over time and in varying environmental settings.

Measurable and observable project manager competencies offer the benefit that training can be customized to bolster strengths and resolve weaknesses. Emphasis on competencies means that the effects of training investment can be measured and evaluated.

2.1.3.3.2. Ensures that the Right Person Gets Promoted.
An advantage of professional project management and its administration by the project organization is that it is easier to measure performance. In its rawest form, the project is either successful or it isn't. In less clear situations, it is straightforward to evaluate the observable and measurable application of best practices by the project manager. The result is that people can be promoted on the basis of performance.

Implementation of a project organization that measures project manager performance, and makes promotions accordingly, eliminates a problem associated with the functionally structured organization. Although the functionally structured organization is highly efficient, one criticism is that it is often difficult to measure the performance of individuals. For example, how does one identify the best accountant in a department of a thousand accountants? Or how does one determine who is the best engineer in a group of a thousand engineers? The response of critics is that the process is difficult. They say that sometimes people are promoted on the basis of other qualities (e.g., networking, personality, etc.) rather than performance. For example, one survey asked respondents "Have you ever worked for or with a manager that you felt was incompetent?" "Yes" answers were received from approximately 90% of respondents. When the researcher delved further it was often determined that the individual in question was in a job where performance measurements were unclear. The survey pointed out the need for performance based measurements such as used in project management when making promotion decisions.

2.1.3.3.3. Increases the Asset Value of the Project Manager.
Professional project management is based upon a solid research and practitioner foundation that the project manager is the key component in achieving project success. Training of project management best practices and competencies makes performance of project managers more predictable. Project managers can be sent to differing project and cultural situations with a higher degree of confidence they will perform their duties in an exemplary manner. In the final analysis, the outcome of professional project management is that the project management professional will consistently outperform the nonprofessional. The result of the process is that the output of the project organization and associated individuals is much higher related to the cost.

2.1.3.4. Benefits to Customers and Stakeholders

Initiation of a project management group improves the relationship between the organization and customers and other stakeholders. The image is presented that the approach to project management is more professional. Adoption of best practices and competencies increases the predictability of successful project completion. The risk of the press reporting project failures and the subsequent embarrassment is minimized. As a result stakeholder morale as well as public image is heightened and stock values remain high. As a case in point, the "Project Value" survey conducted by the *Best Practices Report* determined that 32% of respondents felt that improved customer satisfaction was the major benefit of professional project management.

2.1.3.4.1. Presents the Image of Professionalism and Competence and Predictability. Benchmarkers believe that image is important—particularly when the viewer's perception is that the image reflects reality. In project management it is difficult if not impossible to present the image of professionalism without knowing the language of professional project management and being familiar with its methodologies, best practices, and tools. These competencies are soon apparent to potential customers and other stakeholders. A project management organization and its associated professional project managers clarify the image and provide visible confirmation of proficiency for clients and stakeholders. The impact can be powerful. One benchmarker received a $200 million contract because the customer judged their structured approach to be more professional than the ad hoc manner of the competitors.

2.1.3.4.2. Reduces Risk of Public Embarrassment. No one likes their failures displayed to the public, fellow employees, competitors, and other stakeholders. Failed projects have a ripple effect through an entire organization and in the minds of employees and stakeholders. Often they negatively impact profits and stock values. The appearance of a lack of competence and professionalism makes it more difficult to win bids and generate business. A project organization with its structured approach to project management maximizes the probability of successful project goal achievement across the entire portfolio of projects.

2.1.3.4.3. Supports High Stock Valuations. Ingram's research found a direct linkage between the failure of large functional organizations to perform projects and a reduction in stock valuation. In his study, seven firms lost an average of 59% of their stock value during the time the projects were underway and receiving adverse publicity. Stock price declines ranged from 91 to 20%. Ingram concluded that when functionally organized firms have major losses it is often the result of projects. Conducting their day-to-day business rarely results in major surprises. Implementation of a project management organization and its resultant improved project management performance can be inferred to maintain higher stock values.

2.1.4. Create a Center of Project Excellence

Once the concept of the project organization is approved, the next step is to define the structure of the group. In companies with existing project management cultures in specific functional areas, the project organization can be expanded from the already successful group. Often it is a matter of receiving executive blessing to make the group multifunctional in nature and to have it report to a higher level of authority.

Usually there is no such existing organization upon which to build. In these cases, there is general agreement by benchmarking participants that the easiest and most immediately effective organizational format for the project office is the implementation of a center of excellence. The center of excellence is charged with promoting the value of project management to other areas of the organization. It should be positioned to make project management a true multifunctional team format. It can organize and structure the training for project managers as well as develop standardized methodologies, procedures and templates.

2.1.4.1. Communicate the Benefits of Project Management to Other Enterprise Areas

Although the project group might have received the blessing of senior management, it doesn't mean it is accepted by *all* senior managers. In practice, benchmarkers report that the project group is often greeted with skepticism and outright opposition. As the project group accumulates successes, resistance may increase. Often the challenge comes from higher levels in the organization. Non–project management based departments also sometimes fail to see how the project management group will provide benefits for their specific functional area. Sometimes the project management emphasis is viewed as being "just another management fad." In organizations that are emphasizing cost cutting and downsizing, addition of a project management group is often in direct conflict.

As a result of these obstacles, interviewees judge that one of the first tasks of a new project group is to communicate the value of project management to other areas in the organization. Seventy-one percent of benchmark respondents report that they do, in fact, *formally* and aggressively communicate the value added benefits of the project organization. Benchmarkers emphasize that the process should not be a self-serving approach, but should inform others of the ways the project group can help each functional organization achieve its objectives and be more successful.

Standard: **Best practices project organizations proactively communicate the benefits of the project organization.**

Benchmarkers suggest that when considering the implementation of a project organization the core issue is *change management*. People throughout the organization should recognize the need for change before they can be expected to provide full support. The process can be encouraged by conducting a stakeholder analysis and then communicating the value added benefits resulting from the project organization.

Stakeholder analysis focuses on identifying the project related needs of each stakeholder group as well as key role models in each group. Benchmarkers who have been through the process report that attention is usually given to stakeholder *groups*. However, they emphasize that the stakeholder analysis should also look to the needs of specific *individuals*. Sometimes these are the best and most successful people in the organization. Peers and associates generally respect their views. Benchmarkers advise that the objections of role models should be considered carefully. Sometimes there are good reasons why the change should not be made. It might not be needed or the approach being proposed should be modified.

After analyzing stakeholder needs, development of a communications program can stress *value added* for each of the enterprise units as well as the specific key influencing individuals. Various promotion methods and tools are used. About one-half of benchmarkers surveyed rely solely on internal public relations, while others utilize outside alternatives as well. One-half prepared a formal business plan or proposal that was presented to upper management.

Of the project organizations utilizing external communications sources, about one-quarter relied on articles in nonorganization publications (i.e., professional journals, local newspapers) and one-quarter used outside speakers. Some project organization developers attended professional conferences, made speeches, and participated in professional associations such as the Project Management Institute. It was judged that the external participation activities provided validation and credibility for the employees associated with the project organization efforts. A summary of specific communications methods and tools reported by Benchmarking Forum attendees is as follows:

- **Executive kickoff initiative.** In organizations with senior level commitment, one of the most effective ways to gain support is through an announcement or presentation by the CEO. In one benchmarking organization, the CEO announced, "We are a project managed company."
- **Written reports.** A business plan, white paper, goals paper, or other similar formal presentation serves to formally communicate with senior management. Many organizations include a process for reviewing performance and investigating new ideas and concepts. The formal paper or proposal approach works well in these situations.
- **Internal and external articles.** Articles are one of the most effective means of gaining credibility. Professional or gifted writers from within the group develop articles about the project organization's achievements and strategic approach. These are published in internal publications and external newspapers, magazines, and journals.
- **Outside expert.** Benchmarkers report favorable results from bringing in an outside expert. The individual could be an author of project management books or a respected consultant. Benchmarkers report that outsiders tend to have higher credibility in some organizational cultures than using an insider. The expert makes presentations and talks with people in management. For example, one benchmark organization hired a consultant that gave a two hour presentation and briefing to senior management as well as a general presentation for all other managers. The briefings were followed with two days of training for the project teams.
- **Training.** Best practice project organizations offer formal and informal training to communicate the benefits of project management to other functional areas. These are typically four hour to one day presentations.
- **Web pages.** Organizations develop web pages that describe the project organization and present benefits and other information.
- **Brown bag lunch meetings.** One effective method is to in-

vite people to share their lunch breaks with the project team. Bag lunches are purchased and disbursed. An informal presentation and question/answer approach is taken.

- **Internal and external brochures.** Brochures can describe the project group's achievements, features, and advantages. These are sent to other managers in the organization as well as to external stakeholders.
- **Videos.** Videos have the advantage that they can show the achievements of the group in an entertaining, interesting, controlled, and visual format.
- **Town meetings.** Meetings are held with open invitations for all interested employees to attend. Presentations are made about the project organization and its goals and benefits. People can ask questions, raise concerns, and give their opinions and input.
- **Surveys.** Surveys can be conducted of stakeholders to identify their needs. The survey process provides valuable information plus gives a broad range of managers a medium to provide input into the project organization implementation process.
- **Personal visits to key stakeholders.** Benchmarkers say that one of the best ways to communicate the value of the project organization is to identify specific key individuals and pay them a personal visit. The intent of the visits is the build a cohesive relationship and to encourage the individual to use their role model position to communicate a positive view to others.

Project organizations that actively communicate the benefits of the group report that the efforts are generally successful. The resulting role of the project group is expanded, and the group's stature and positioning in the organization is elevated.

2.1.4.2. Make Project Management Multifunctional

One of the strongest and empirically robust links to project goal achievement is that the project groups be multifunctional in composition. Immediate positive results can be expected by including on the project team representatives from all functional areas in which the project will interface. For example, a new product development group might include design engineer, marketing, production, and finance personnel. The multifunctional team members work simultaneously, in a parallel effort, on each of the functional aspects of the project. Replaced is the traditional sequential approach of functional organizations where each functional area performs its activity and then passes the project to the next functional area.

Participants in the benchmarking forum report a high degree of success in implementing the multifunctional approach. About two-thirds of benchmarking project organizations report responsibility for more than one functional area with nearly half recording responsibility for *all* functional areas for which their projects interface.

Benchmarkers also relate that host organizations are sometimes reluctant to give the project groups complete authority over money and people. Research as supported by practitioner experience supports the conclusion that the probability of project success is increased when the project organization has responsibility for all financial activities (e.g., the budget and day-to-day accounting) as well as hiring, firing, compensation, and training.

2.1.4.3. Train Project Managers

Additional benefits can be attained for the organization by initiating a competency based training program. These consist of the off-the-shelf or customized training programs offered or developed by universities, the Project Management Institute, and consulting organizations. Taught are the methodologies of project management as well as other components of the project body of knowledge such as leadership, contracts, international aspects, finance and accounting, and software applications. All are skills and best practices that can be immediately applied to the job. As a result there are immediate project management performance improvements.

2.1.4.4. Develop Standardized Methodologies, Procedures, and Templates

Benchmarkers report that the third area of emphasis for the new project organization is the development of standardized methodologies including the selection of tools and templates. Many companies dedicate considerable time to this process, although the easiest approach is to utilize any of the off-the-shelf packages. In the appendix of this book are the major documents used in typical standardized methodology packages.

2.2. TIME REQUIRED TO ESTABLISH THE PROJECT ORGANIZATION

Benchmarkers voice the opinion that it is possible to initiate multifunctional teams, start a training program, and develop a methodology quickly, although generally about two to three years are required to have the organizational culture accept the office and its activities. A typical development schedule is to immediately begin implementing multifunctional project teams and the training program. The first year is also spent acquiring and implementing the standard methodology and associated procedures and templates. During this time, project management as a concept is promoted throughout the organization. The second year is dedicated to encouraging acceptance of project management as part of the organizational culture and communicating successful results. In year three of a typical project organization implementation plan, the organization has likely started recognizing the value of project management as a result of positive experience with the discipline. During this process the success of the project organization will be evidenced as functional departments ask for increasing levels of support and involvement. Benchmarkers agree that there is generally a happy ending as project successes accumulate. Usually professional project management is fully accepted after three years in most organizations.

3

Project Organization Structures

3.1. BUILD AN EFFECTIVE AND EFFICIENT ORGANIZATIONAL STRUCTURE

Structure and placement of the project management group within the host organization are critical. As early as 1916, Henry Fayol, in his book *General and Industrial Management,* emphasized the importance of structure, or specifically, creating the structure and lines of authority and responsibility for the organization. Best practices project organizations are structured for goal achievement. Bureaucracy is minimized, staffs are small, there are few layers of management, spans of control are broad, and information flows freely.

> **Standard:** Best practices project organizations design the project group structure to be compatible with and support the organizational strategy and structure.

3.2. SENIOR LEVEL RESPONSIBILITY

Benchmarkers agree that the highest probability of success is attained by having the project organization report to a multifunctional senior level executive or group. Clearly a core best practice is for the project organization to be placed at a high level in the organization. Being placed at a high level gives the project group formal authority to achieve its objectives as well as making it more intimately familiar with organizational strategies and tactical plans. It positions the project organizations to have freer access to resources. The project organization needs sufficient stature within the host organization to engage in political maneuvering and power struggles between departments.

Standard: **Best practices project organizations report to a multifunctional executive or executive committee.**

References: Ancona and Caldwell, 1992b; Zangwill, 1992.

3.3. PROJECT ORGANIZATION TYPES REPORTED BY BENCHMARK RESPONDENTS

Participants in the benchmarking forum were surveyed and interviewed to define the various types of project organizations and to determine how each of the resultant models fit different corporate cultures and organizational structures. Figure 3.1 shows the generalized depiction of project organizations as detailed by forum attendees.

Organizations in the benchmarking forum tend to have four basic types of project group structures: the Center of Excellence provides project leadership, promotes the value of project management, trains, and implements methodology; a Project Support Office provides consulting services to project groups throughout the organization; the Program Management Office manages project programs and the portfolios of projects; and, finally, the pure Project Office manages a single project. It is emphasized that of the 60 organizations in the benchmarking forum no two reported exactly the same type of project group structure. Many overlapped and combined various functions and activities. Also, the figure by inference shows a hierarchy with

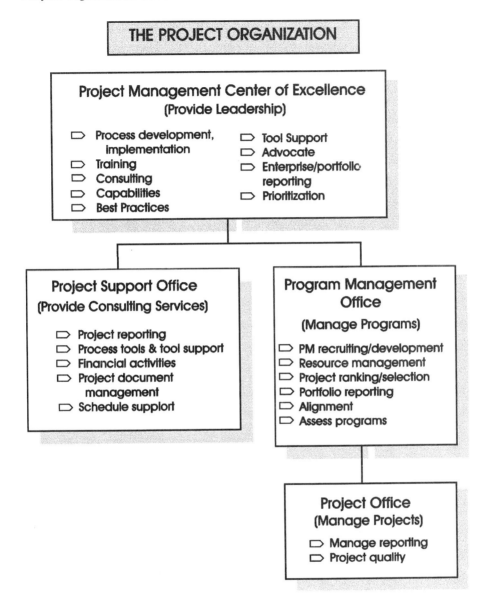

FIGURE 3.1

the Center of Excellence at the top. This is not always the case. Although many organizations use the Center of Excellence in combination with other types of project offices, there were numerous situations where the Project Support, Project Management, or Project Office was found as an independent unit. The details of each specific approach follow.

3.3.1. Project Management Center of Excellence

The project management Center of Excellence provides administrative and leadership functions. Its strategic focus is aimed at improving the capability of the corporation in managing projects. It serves an advocacy and ambassador role. The Center of Excellence is a corporate rather than a divisional resource and is funded as an overhead expense. The center of excellence provides executive oversight of projects from a portfolio, quality, and stakeholder view. It does not manage projects nor provide project managers to consult with specific projects, although it is often combined with the Project Support and Program Management functions. More companies in the Benchmarking Forum (40%) report having a Center of Excellence than any other form of project organization.

The Center of Excellence approach seems particularly appropriate for large companies with diverse divisions and global operations. Typical examples are Fortune 100 companies in the telecommunications industry, personal computer and software manufacturers, high tech products manufacturers, and various global manufacturing organizations. Several of the companies embracing the Center of Excellence structure have thousands of project managers in their organization. For example, one company has 10,000 project managers and numerous project support offices distributed throughout the world. Specific activities performed by the Center of Excellence and the percentage of companies reporting the activities are as follows (see Table 3.1).

TABLE 3.1 Project Management Center of Excellence Detail and Weighting of Activities

Type of activity	Number reporting	Percentage reporting
Processes	12	19
Training	10	16
Consulting	8	13
Capabilities	8	13
Common approaches	7	11
Advocacy	6	9
Best practices	5	8
Enterprise/portfolio reporting	3	5
Establishing priorities	4	6
Total	63	100

Processes. The most commonly mentioned activities performed by the Center of Excellence involves process development, implementation, and maintenance. The processes involve project management methodologies, organizational models, solution practices, and ownership issues. Common communications approaches are particularly emphasized in large globally diverse organizations.

The Center of Excellence is particularly suitable where the organization is using virtual project management offices and there are multiple project organizations dispersed globally. These organizations report increasing use of web pages and software tools such as Lotus Notes. Several companies are initiating global information systems for project management.

As an example of the importance of process management by the project organization, one Center of Excellence in a large insurance company reports that they have 17 offices around the globe and 60 corporate wide projects underway. They utilize project staff and line managers from the various functional and global offices to execute the projects. Consistent processes ensure effective communications and predictability in the manner projects will be performed.

Training. The Center of Excellence emphasizes training. A major function of the Center of Excellence is to raise the organization's project management competence across all company projects, facilities, and business units. The process includes evaluating performance and comparing it with specific competencies.

The training function includes the development of curriculums and training programs for project managers and teams and facilitation of training that nurtures soft skills as well as technical ones. Mentoring on a one-to-one basis is included. As a tangible measurement of training accomplishment, several companies require PMP certification for senior project managers.

Some companies in the Benchmarking Forum send thousands of project managers through training programs. Several of the organizations have sent over a thousand people through project management training. One participating company is training 15,000+ managers in project management skills.

Consulting. Personnel from the center of excellence consult and counsel project managers, other functional areas, groups, and departments that manage their own projects. The consulting function includes mentoring and monitoring projects.

Project management capabilities. There is a broad base of activities that include development of career paths, assessment of organizational skills, definition of project manager competencies, and development of performance metrics. As one participant stated, "Their role is to improve organizational competence for managing projects from *above, across* projects and *within* the organization."

Common approaches. A key activity of the Center of Excellence is to insure consistency across the organization. It includes definition of common methodology, tools, and training. The function has greatest prominence in large companies with diverse geographical operations.

Advocacy. Centers of Excellence consider it important to promote project management to senior management and other functional areas. The office serves as an advocate for professional project management. The process involves finding and supporting project management champions. They serve as the "owners of the mission" to propagate project management. Benefits of specific projects and project management in general are communicated throughout the organization. One global participant states their advocacy role in terms of three goals: (a) create a core competence for project management across the company, (b) gain commitment from business units to have a project management implementation plan, and (c) organize a corporate wide council for networking.

Best practices. The group evaluates other organizations and applies knowledge gained to their own organization. Participation in professional benchmarking; research of best practices, competencies, and performance standards; and analysis of lessons learned provide information for improvement.

Enterprise/portfolio reporting. A few Centers of Excellence are controlling and reporting the status of the enterprise project portfolio.

Establishing priorities. Some Centers of Excellence coordinate project portfolios, analyze requests for project management, assist in cross-integration of multiple department projects, and align projects to strategic goals and strategies.

3.3.2. Project Support Office

The Project Support Office provides administrative and consulting support to project managers. The Project Support Office does not directly manage projects or supervise project managers. It is funded by the project groups and usually billed directly. It includes functions that support the project manager's role and the activities of the project. The benchmarking companies report that the project support office is nearly as popular as the Center of Excellence with 34% reporting its use. Senior management in organizations evaluated has responded favorably to the Project Support Office structure because it results in better predictability of project performance. Note that several companies use both the Center of Excellence and a Project Support Office.

The Project Support Office serves as a central resource for project management skills and consulting services. The concept is most appropriate in large companies where there are broad project management needs and varying centers of expertise and specialization.

One company had evolved to a project support office concept after having tried the Center of Excellence approach. They desired a more "in the trenches" approach. In the view of the participants, it was felt the Project Support Office provides more value to projects and other corporate stakeholders. There is more influence on project outcomes if the project is approached from a consulting viewpoint.

Benchmarkers using the Project Support Office concept report that there tend to be many relationship based issues and that accountability is sometimes a problem. The division of accountability, responsibility, and authority between the Project Support Office consultant, project teams, and functional areas need to be defined. Deliverables for each group should be clearly identified. One of the largest Project Support Office organizations professes that they accept no responsibility for the success of the project—it is strictly under the control of the functional area performing the work.

Companies with Project Support Offices stress the need to fully understand the value of professional project management. The services of the project consultant are typically charged to the user department. Companies experiencing the highest success rates with the Project Support Office concept are those that track project costs closely as well as the cost of professional project management. Cost of professional project management typically runs from 3 to 15% of total project costs in for-profit organizations reporting (see Table 3.2).

Project reporting. Project Support Offices provide central reporting and a project information repository or knowledge library. They monitor, issue, prepare, and manage reports. A primary function is to transfer knowledge from one project to the next and increase overall efficiency of all projects. They

TABLE 3.2 Project Support Office Detail and Weighting of Activities

Type of activity	Number reporting	Percentage reporting
Project reporting	18	39
Process tool selection and support	10	21
Project portfolio	6	13
Financial activities	5	11
Project document management	2	4
Schedule reporting	1	2
Miscellaneous project consulting	2	4
Training	2	4
Project oversight	1	2
Total	46	100

communicate project status, progress, and variances; track costs; make staffing plans; and identify deliverables.

Process tool selection and support. Tool expertise, support, and maintenance are developed. The Project Support Office assists with data entry, the building of work breakdown structures, project schedules, and earned value project performance measurement usage. Encouraged is the utilization of standardized methodologies and project management approaches.

Project portfolio administration. Project Support Offices administer change control, scope management, and risk evaluation and mitigation. They support and sometimes conduct project audits. Managed are portfolio and program issues.

Financial activities. Administered are accounting (e.g., payables, receivables, and billing), as well as costing and pricing support and subcontracts.

Project document management. Project data such as the work breakdown structure, estimates, and schedules are managed. These are coordinated and distributed as needed. The project knowledge library is maintained.

Schedule reporting. Scheduling services are provided.

Miscellaneous consulting. The Project Support Office occasionally will send out project managers to work with specific projects. The project managers tend to be project management rather than domain or project specific technology specialists. Benchmarkers warn that it is always necessary to clearly define accountability in these situations.

Training. Training conducted in association with the Project Support Office tends to be specific to project related problems. One company reports that they use "just in time training." They go directly from problem discussion and resolution in the class room to application in the work place.

Project oversight. One company reports that their Project Support Office serves an oversight function. They review project performance and audit processes.

3.3.3. Program Management Office

Program Management Offices work well where the company has corporate wide directives involving multiple efforts. A *program* is a collection or portfolio of projects with a common vision. The Program Management Office looks at strategic as well as tactical issues. The Program Management Office serves as a focal point in each business unit as the project management organizers.

The office is funded directly from projects and programs. Program Management Offices are reported by 16% of Forum participants. Program Management Offices are often used in conjunction with Centers of Excellence as well as Project Support Offices.

Most program management offices are at high levels in the organization. The focus of Program Management Offices is on high visibility, cross-functional projects involving multiple corporate areas. The manager of the office may be at vice-presidential level in the organization. In one benchmarking company the manager of the Program Management Office is given the title of President and reports directly to the Chief Executive Officer of the organization. The Program Management Office manages, delivers, and is accountable for the project managers, programs, and program portfolios. Specifically, the Program Office has profit and loss responsibility. The Program Management Office is responsible for administration and strategic and tactical control of all project management activities. The Program Management Office typically has a dotted line responsibility to the functional organizations.

Some Program Management Offices have met resistance from entrenched functional organizations. As the Program Management Office group acquires increasing successes it becomes more powerful and threatening to traditional groups. The conclusion of benchmarkers is that politics and territorial issues are often involved in working with the functional leaders.

The Program Management Office focuses more on portfolio management more than any of the other organizational forms. Activities

TABLE 3.3 Program Management Office Detail and Weighting of Activities

Type of activity	Number reporting	Percentage reporting
Recruiting and developing project managers	7	19
Resource assignment/management	4	11
Project selection and prioritization	5	14
Portfolio management	4	11
Strategy alignment	7	19
Program assessment	4	11
Communication	1	3
Miscellaneous	4	11
Total	37	100

reported include project selection and strategy alignment as well as program assessment. Other aspects of the office's functions are project manager recruiting and development as well as communication. See Table 3.3.

Recruiting and developing project managers. In common with other project office organizations, a considerable amount of time is spent developing project manager career paths, providing performance and feedback reviews, and recruiting.

Resource assignment/management. The portfolio management aspect includes the assignment of resources, workforce planning, provision of the project manager pool, and provision of backup project managers.

Project selection and prioritization. The office evaluates risk and ensures that projects selected are prioritized according to senior management's communication of their strategic roles within the organization.

Portfolio management. By definition, the Program Management Office coordinates multiple projects and programs. A primary activity for effective project portfolio management is to first conduct an inventory of existing projects and services. From this a ranking and evaluation system can be administered and the quality of the portfolio gradually refined.

Strategy alignment. The office provides cross-functional linkage between the participating organizational bodies. It sets the group mission and integrates that mission with company core value statements. It adopts a strategic plan for enterprise projects and sets measurable strategic plans.

Program assessment. Metrics for project evaluation are developed and administered. Early warning indicators are used to identify projects in trouble and to spotlight and prevent problems in their initial stages of development. Included are risk assessment checklists and templates. Results of all projects are rolled up, evaluated, and reported. Performance and project contributions to the organization are measured and communicated to stakeholders.

Communication. Tools are provided to improve communications between projects and diverse geographical areas. Web sites are administered, which are becoming the standard approach for project organization in multiple geographical sites. The group provides the tools, database, and tools support and mentors project managers in other internal organizations. Basically the office ensures that the project teams work together toward common objectives.

Miscellaneous. Other activities include provision of smoke jumpers to save projects in trouble.

3.3.4. Project Office

The Project Office controls and manages one project. A typical example would be the trailer on a construction site. The project manager and associated personnel and support materials are contained at the Project Office. The Project Office typically will plan, control, and implement a single project. Only 10% of respondents report that the Project Office is part of their project organization. Most companies indicate that specific projects are controlled and implemented within the framework on functional units.

The Project Office manages an individual project. It is responsible for the project deliverables, is accountable for performance, and provides leadership for the project team. The office sets and maintains quality standards, methods, and processes and reports project status to stakeholders. See Table 3.4.

Note: **The standards and best practices and competencies of** *Project Offices are covered in the companion volume* **The Superior Project Manager: Global Competency Standards and Best Practices.** *They are the same as those detailed for project managers who manage a single project.*

TABLE 3.4 Project Office Detail and Weighting of Activities

Type of activity	Number reporting	Percentage reporting
Project management	8	67
Quality assurance	3	25
Miscellaneous	1	8
Total	12	100

3.4. CURRENT REPORTING STRUCTURE OF BENCHMARKING PARTICIPANTS

Despite the trend for project groups to be cross-functional in nature, 72% of benchmarking organizations report that their project organizations fall within the engineering department. About one-third of these report to a vice-president with the remainder reporting to a manager. Three of the groups report to the president or chief executive officer of the company.

3.5. MULTIPLE PROJECT OFFICE ORGANIZATIONS

Several large organizations in the Benchmarking Forum have multiple project "offices" or organizations. In most cases the organizations have evolved in response to varying needs and distinct problems of diverse functional and geographic enterprise units. Often there is significant overlap of activities. The benchmarkers say that before adding multiple offices the same decisionmaking process should be applied as when the first office was started. Specifically, the proposed office or project organization should clearly add value to the project management process. In for-profit organizations there should be a clear expected improvement on the bottom line.

Benchmarking organizations say that when multiple project organizations exist there must be formal efforts to minimize overlap of activities and its associated increase in expenses, to share resources, and to align strategies and goals. Where project organizations have spontaneously developed, there should be an analysis to determine if they should be merged into larger units.

Part II

Providing Support for Organizational Strategy

4

Coordinating Goals and Missions

4.1. PROVIDE SUPPORT FOR THE ORGANIZATION'S STRATEGY

Project management practitioners consistently express the view that the ultimate purpose of the *project* organization is to improve the ability of the *host* organization to achieve its goals.

> **Standard:** **The best practices project organization directs all activities to maximize the probability of achieving the host organization's goals.**

The value of attaining unity between organizational and individual project strategy has been documented for thousands of years. In historical military strategy books such as Sun Tzu's *The Art of War* (ca. 500 BC) and Carl Von Clausewitz's *On War* (1850) it is repeatedly stressed that effective strategic and tactical planning treats each battle as a key building block in achieving the organization's goals (e.g., winning the war).

Researcher Sheriff found that setting clear goals improves team alignment with the desires of the host organization and stakeholders. The translation of the host organization's goals into short-term project activities is key to project success. The formalized methodology adopted as a part of the project organization implementation process encourages the team to focus more on the critical success elements as well as the performance factors on which the team will be judged.

Zangwill found that in addition to setting *overall* project organization goals, superior teams establish *specific* process goals that define performance of the individual project teams. Goals force the groups to look ahead, identify potential risks and problems, and plan for flexibility in advance.

4.1.2. Coordinate Goals and Missions Between Project Teams and the Host Organization

4.1.2.1. The Coordination Problem

Benchmarkers relate that a problem encountered by teams located within functional departments is a lack of understanding about how their efforts interface with and support the host organization's goals, objectives, and mission. Without a full understanding of the strategic fit, there is a tendency for the project team to flounder, lack direction, drift, and gradually become diverted from its primary purpose. Multiple research studies show that conflict within the group results when evidence indicates a divergence between the project goals and those of the host organization.

The situation becomes particularly critical when there are large numbers of uncoordinated projects. For effective project portfolio management, the consensus of practitioners is that there should be a direct articulation between corporate and project portfolio strategies. A clear understanding is crucial regarding the relationship between each project with the overall portfolio as well as the organizational vision and mission.

4.1.2.2. Communicating the Vision, Mission, and Goals

Benchmarkers report that the solution to the need for understanding the host organization's vision, mission, and goals is for the project organization to translate and communicate the linkage between the goals and mission of the host organization with those of the individual projects.

> **Standard:** **The best practices project organizations aligns and integrates vision, goals, and plans of individual projects with those of the host organization.**

4.1.2.3. The Project Organization's Goals

Benchmarkers emphasize that before the host organization strategy can be communicated with the teams, there should be a comprehensive understanding of the relationship between the project organization's goals and those of the host organization. Further, it should be clearly understood how the project organization will support the attainment of the host organization's goals through the implementation and support provided for specific projects as well as the overall project portfolio.

4.1.3. Set Project Goals and Deliverables

The attainment of the host organization goals and the related project tactics becomes achievable through restatement in the form of *project organization* goals and deliverables. The goals and deliverables focus the project organization on the critical success factors upon which the performance of the group will be judged. Management maintains focus on portfolio objectives and the tendency to overemphasize high profile projects is reduced. The project organization is encouraged to look ahead, identify potential portfolio risks and problems, and plan for flexibility. Alignment is created. Limited is time-consuming bickering between the project organization, individual project teams, the host organization, and other stakeholders.

> **Standard:** **The best practices project organization set clear, measurable, and observable goals.**

After the project organization's goals related to the host organization are understood, *project specific goals* can be developed. Benchmarkers say that often there is a feeling that the importance of the project is

so obvious that it is known to all. This is usually not the case. Consequently, the communication of the association between host and specific project goals first occurs during the initiation and selection process. The topic is formally addressed in the project charter. A part
of the selection process methodology of some benchmarking project
groups is to *define the strategic context* of each project being evaluated. When setting context, questions are addressed such as: "How
does the project effort fit within the organization's strategy?"; "What
specific organizational goals and objectives will be attained or supported by this project?"; "How will the project interface with and support other initiatives and ongoing work?"

After the project is approved, the charter is typically expanded
to become the project plan. At this point the strategy and mission as
described in the charter can be linked to specific project deliverables,
goals, and activities. Once the project goals are set, all other activities
can be evaluated based upon the manner in which they help achieve
the project and, ultimately, the host organization's goals. Included is
measuring project performance as well as the competence of project
managers. Assurance is maximized that the project plan supports the
host organization's strategy.

4.1.4. Maintain Focus on the Goals

During the execution phase of the project, the project organization is
charged with monitoring the goal achievement progress of the portfolio of projects. The task is critical because practitioner experience as
well as research indicate that maintaining a constant focus on the
goal of the group has a high correlation with goal achievement. For
goals and deliverables of individual projects within the portfolio to
become reality, the superior project organization ensures that goal
focus is stressed to the point it becomes an overshadowing factor in
making daily decisions. Prevented are unjustified scope changes
that alter the original goals as defined by the project charter and
plan.

> **Standard:** **The best practices project organiza-
> tion maintains constant focus on the
> host organization goals as supported
> by the project organization and spe-
> cific project goals.**

References: Toney, 1996; Brown and Eisenhardt, 1995; Zangwill, 1992; Chandy, 1991; Ancona and Caldwell, 1990; Miller et al., 1985; Shaw, 1972; Schmidt and Kochan, 1972; Likert, 1961; Sherif et al., 1961.

4.2. MAKE PROJECT MANAGEMENT A CORE ORGANIZATIONAL COMPETENCY

Numerous organizations recognize the value of professional project management. Of the representatives attending the Benchmarking Forums, approximately 60% say that their organizations identify professional project management as an important element of organizational strategy. Several organizations go a step farther and assert that for the project organization to achieve maximum success, professional project management should be recognized as a *core competency* of the host organization. Core competencies are described by the benchmarkers as a body of actions that are critical to the success of the organization. One best practices organization said that they defined a core competency within their organization as a strategic competency that would not be compromised.

All the benefits previously discussed throughout this book support the conclusion that professional project management is a critical function and should be formally recognized as such. As project organizations accumulate successes, clients and stakeholders increasingly demand higher levels of professionalism in managing projects. In some cases, particularly where projects are revenue producers, it is recognized that professional project management represents increased profit opportunity. One best practices organization emphasized that almost all new business opportunities are projects. The manner in which they are governed has a direct impact on the future success of the organization.

In some organizations, senior management publicly states that professional project management is a core process in the strategic foundation of the organization. One company has stated, "Projects are the basic building blocks of our business."

> **Standard:** **Best practices project organizations make professional project management a core strategic process within the host organization.**

Benchmarkers relate that making project management a core organizational competency is not without its political and cultural aspects. In general, there is acceptance of the need for professional project management as a core competency. However, there is often disagreement about how the objective will be accomplished within the cultural constraints of the organization.

4.3. INTEGRATE PROJECT MANAGEMENT END TO END IN THE VALUE ADDED CHAIN

One outcome of recognizing project management as a core competency is that the project management methodologies become integrated throughout the entire value chain of the organization. In particular, it means involving the project group at the origin of strategy development, at project initiation and selection, and/or at the sales stage.

There is agreement among benchmarking forum participants that the earlier the involvement of project leadership in the project, the higher the probability of project success. Benchmarkers report that early involvement of the project organization improves project flow through the organization. It improves accountability because the project management is part of the early decision process.

Research by Gupta and Wilemon supports the benchmarkers' experiences with hard scientific data. In the Gupta and Wilemon study, 42% of respondents state that early involvement of the multifunctional team is a key element in attaining project success. They say that integration and involvement of the project group at an early stage in the process results in greater shared commitment, clarified stakeholder needs, and better definition of product specifications before large amounts of money are spent. Response time is reduced because future changes are minimized. Cooperation of the functional groups is improved. In the Gupta and Wilemon research, 71% of respondents say that projects that fail to meet time and cost targets are the result of poor definition and understanding of customer requirements and insufficient knowledge by sales people of the product's technology.

> **Standard:**   **The best practices project organization is actively involved in the project management process from inception to termination.**

On complex project proposals some companies send the project manager to visit the customer along with the sales person. The process has insured that the client has better understanding of deliverables, the project process, and the relationships between the stakeholders. Overall project speed, efficiency, and effectiveness improve as a reflection of the clearer definition of project functional and technical specifications. Minimized is the problem of features and outcomes being sold that are outside the scope of the project team to perform.

5

Project Portfolio Management: Inventory, Initiation, and Selection

A *portfolio* of projects includes all the projects monitored and tracked, projects for which consulting is provided and projects actively managed by the project organization. It is the broad group of projects over which the project organization exerts coordinated and strategic control.

Many project organizations initially embrace portfolio management as an *informal* response to the increasing number of projects under their responsibility. The project organization often begins managing the group of projects as a portfolio because no one else is doing it and they are in an advantageous position to accomplish the task. As positive results from the process accumulate, the project organization normally adopts a more formal approach.

The progression of most project organizations in the benchmarking forum is to begin with the development of a database and inventory of existing projects. The portfolio of projects is then ranked by appropriate factors such as strategic importance, financial return, and risk. Then defined are the governance process, the type of projects included in the portfolio to be managed, project ownership, and cultural relationships with senior management and the functional departments and global divisions. As the day-to-day management of the project portfolio evolves, the project organization defines the processes to be applied for new project initiation and selection, control and management, and termination of existing projects.

Benchmarkers report that before project portfolio management, many organizations had a few priority projects. There are now many; and the many are managed to optimize their organization wide value in achieving strategic goals. In a way, the large mass of projects is like a group of players in a team sporting event. Most would agree that the players will be more effective as a coordinated team than as a random assemblage of individuals. From a business perspective, the manner in which the project organization approaches the portfolio of projects is the same as the manner in which financial people handle their portfolio of investments.

The benefits of project portfolio management are many. For corporations, more profit can be earned at the same time overall risk is reduced. For nonprofit enterprises, the probability of goal achievement is increased with the same corresponding reduction in risk. The entire portfolio of projects can be focused on strategic issues. The approach is particularly applicable to global operations where projects and project team members are dispersed throughout the world.

Project portfolio management balances the value and needs of individual projects with those of the entire organization. It presents a clear picture of the impact on resources and schedules of adding each new project to the portfolio. The overall portfolio becomes more efficient. Duplicate, marginal, and losing projects can be eliminated or spotlighted for continuing scrutiny. Oversight of scope changes maintains focus on the attainment of strategic goals. The evaluation and selection process can eliminate spontaneous projects. Budgeting can be more uniformly administered. Projects can receive funding as a reflection of their profit, risk, and strategic values.

With project portfolio management, the performance of single projects related to the portfolio or the average project, becomes easier to measure. All of the measurement metrics used by the project organization can be aggregated over a period of time and across numerous projects. A high level summary of all projects can be provided with data regarding performance, strategic effectiveness, and any pertinent specifics desired such as personnel utilized and funds expended and required. The rolling up of portfolio metrics gives the project organization capability to evaluate changes and to set goals for future projects.

For projects that don't fit neatly into traditional organization charts, such as information systems, software, and internal process and cross-functional systems improvements, the portfolio approach helps define overall organizational needs. Ownership and responsibility of project deliverables and outcomes is clearer. There is more control over the deliverables and budgets of these sometimes vague and nebulous projects.

> **Standard:** **The best practices project management organization managers projects as a coordinated portfolio.**

5.1. PORTFOLIO GOVERNANCE

The management of most portfolios in the benchmarking forum report to an executive committee oversight or governance board. The boards are typically composed of senior level executives. In one large company, the portfolio manager reports directly to the president of the company. In all cases, the portfolio manager provides detailed communications about portfolio status as well as individual projects at the board meetings.

5.2. PROJECT OWNERSHIP AND ACCOUNTABILITY

Benchmark participants stress that project ownership is frequently a matter of confusion and vagueness in large functionally and globally diverse organizations. Ultimate responsibility and accountability becomes even less clear when the project becomes part of the enterprise portfolio. The resolution of the issue is important to project goal achievement. Gupta and Wilemon's study of failed projects finds that when accountability is unclear, project deficiencies result such as insufficient monitoring of project progress, poorly applied control systems, undefined and conflicting roles, and complex team organizational structures. The Standish Group provided support for the conclusion with their determination that definition of project ownership is a key project success factor.

Benchmarkers say the ownership question is easy to state: "Who is responsible if the project is a failure?" or "Who is ultimately accountable for the success of the project?" The answer to these questions is unclear in many organizations. There is widespread opinion that ownership rests with the project manager and team, particularly when working with vague and nebulous projects such as software and information systems. The view is heightened when the originating department or manager is unsure about exactly what they need and how the service will be delivered. In these cases there is a tendency to shift the burden of ultimate performance to the project team.

Benchmarkers say that the "widespread opinion" is incorrect and that final project accountability and responsibility rests with the client, sponsor, or originating owner of the project—not the project manager and team. If a functional or global department or division originates the project, they are the owners and assume ultimate responsibility. If it is an enterprise wide initiative, senior management is the owner and responsible group. An example used by forum participants to demonstrate their judgment relates to building a new home on the bank of a river. The question could be asked, "Who is responsible for the risk of flooding—the home owner or the general contractor who is constructing the house?" Most would agree that the owner of the home is the person accountable and responsible.

It is the contractor's (project manager's) role to *communicate* risks, their probability of occurrence, potential magnitude and response plan to the homeowner. The role of project manager and team is to satisfy the needs and project plan as approved by the client or sponsor. In cases where the customer is unsure of what they want or will be getting, it is the project manager's role to help define their needs and translate those needs into the detailed project plan.

Benchmarkers say that often the division of responsibility is less clear than presented in the home building example. They suggest that the situation is more similar to hiring of expert professionals such as an attorney or CPA. In many cases the professionals themselves infer or even communicate that they are taking charge. The presumption is that they are accepting ownership, accountability, and responsibility. Experienced project organization practitioners would caution that the client or sponsor of the project is ultimately responsible for its success—not the attorney or CPA. The professional is an expert advisor and counselor.

The danger of having the project manager assume ownership is that even when the project is implemented on time and within budget, it can still be considered a failure if it doesn't meet the user's needs. Consequently, having the client or project owner know that they are ultimately responsible for project success results in a higher degree of involvement in all the aspects of project planning and execution. A clear acknowledgement of ownership helps resolve glitches and organizational bottlenecks. The accountability aspect of project ownership reduces the inclination of project stakeholders to degenerate into finger pointing, name calling, and confusion.

The benchmarking participants emphasize that project ownership, responsibility, and accountability should be defined clearly and early in the project process. For projects that are vague and nebulous and span global and functional areas, the question can often be answered by determining who or which group provides project funding. For projects where *no one* wishes to accept ownership, benchmarkers would suggest that the question should be asked whether the project is viable and should be performed.

One benchmarking representative from a global company related their experience that a key to project success was when there is an enthusiastic sponsor for the project who funds the project and also wants to remain in control. Another best practices project organization has a requirement that a single individual must be assigned overall ownership and responsibility for every project in the portfolio.

> **Standard:** **Best practices project organizations clearly define and communicate that ultimate responsibility and accountability for the project rests with the project client or owner.**

5.3. CULTURAL PROBLEMS

Benchmarking representatives relate that the portfolio implementation process is not without its political aspects. As the project organization investigates large projects spread across the organization, it sometimes becomes apparent that flaws exist. The weaknesses of each functional area become visible. These are often at high levels and can be threatening to the people there.

For many organizations, the strategic management of the project portfolio represents a major shift in business direction and culture. It is a transfer of power and resources away from the functional and global units to a more project based form of governance. For functional and global units that have traditionally controlled the majority of resources for projects affecting their area, removal of any degree of financial responsibility is sometimes viewed as a reduction of status and formal authority.

Benchmarkers agree that for portfolio project management to be effective there should be full senior level support and agreement between functional and global areas about which projects are subject to portfolio management. All stakeholders should fully understand the strategic benefits to the organization as well as their own areas of responsibility.

5.4. PROJECTS INCLUDED IN THE PORTFOLIO

As the project organization accumulates more and more projects to manage and monitor or for which to provide consulting services, a decision must be made about which projects to include in the project portfolio. Although there are a broad variety of approaches among benchmarking organizations, there are two generally accepted views. First are those who include *all* formally approved projects. Second are those who include *almost all* formally approved projects. For project organizations that accept almost all projects, there is usually an arbitrary minimum size such as undertaking projects with durations over two weeks.

All would agree that cross-functional projects are prime candidates for the portfolio. Also included are large projects including those that impact only one department or global division. Some might suggest that the functional or global department is better positioned to perform their own project activities than the project organization. Benchmarkers testify that in practice, the project organization is more effective because frequently the functional or global personnel are dedicated to other activities and the time-consuming monitoring and execution of projects is a distraction from day-to-day affairs.

Some benchmarking organizations exclude small projects from the portfolio. One reason given is that resources are limited. Even though the small projects may already be funded by the sponsoring group, other resource limitations such as project manager and other personnel availability can be factors. As a result priorities are established.

Benchmarkers who reject the smaller projects also say that when the project organization accepts all projects submitted, it effectively results in a project priority system based upon "who knocks on the door." They say that that projects accepted by the project organization, including the smallest, should be subject to a consistent selection, evaluation and ranking process. As support, the benchmarkers relate examples where projects have been initiated by the functional areas without passing through a rigorous approval process. A related problem encountered by benchmarkers is that functional organizations sometimes hand off marginal and failing projects for which they are unwilling to terminate. The benchmarkers assert that the rejection process resolves these predicaments as soon as possible before sizable funds and resources have been allocated and spent.

Benchmarkers also say that the open door policy results in a tendency to become overwhelmed with small projects. In one company, the inventory process determined that inordinate amounts of time were dedicated to small projects. Senior management judged that they were a major consumer of resources. As a result, the project group worked to reduce the number of projects in process rather than encourage acceptance of more.

The benchmarking group that accepts all approved projects agrees that there is constant temptation to reject smaller projects. However, they feel that there are good reasons for being more receptive. By accepting the smaller projects, better organizational control is maintained. A view is provided of the composition of the *total* organizational portfolio. The need and usage of resources can be more easily documented. The benchmarkers referred to the previous example about the company where small projects were assuming excessive resources. They say that had these smaller projects been excluded from the portfolio, the information would not have been available for decisionmaking. By being included in the portfolio, small projects are inventoried and prioritized in the same manner as larger projects. They can be subjected to a pragmatic evaluation process.

There is also concern about what happens to the projects that have been rejected and remain outside the portfolio monitoring and control system. At least in theory, all projects proposed were developed in response to a perceived need. After rejection by the project organization, the need will continue and the project will likely still be executed. The experience of benchmarkers is that when numerous small projects are rejected, the functional or global organization is inclined to start a mini–project organization that is a shadow or duplicate of its enterprise counterpart. As an example, benchmark attendees report numerous organizations that have multiple information technology departments located throughout the organization. Often the functional or global groups hire their own programmers and contract with outside vendors. The duplication of services results in higher project costs, lower levels of project management expertise, and resultant lower standards of performance.

Benchmarkers say that there is also an image aspect to accepting all valid projects. Most people don't like rejection, particularly when they perceive themselves to be a potential client or supporter.

Benchmarking participants relate that it is difficult to tell people that their project is too small or doesn't meet the minimum requirements for acceptance into the portfolio. If rejections are received repeatedly with small projects, there is a tendency for people to be reluctant to return with their larger projects. Accepting all valid projects presents a positive image that encourages people to embrace the project organization.

5.5. THE PORTFOLIO MANAGEMENT PROCESS

As the value of portfolio project management is recognized throughout an organization, the tendency is to approach the task in an increasingly structured and professional manner. The knowledge model upon which the project portfolio management is patterned is the general financial subject area of capital management. Project portfolio and capital management have many similarities. Both projects and capital expenditures involve large amounts of money. The direct and indirect impacts of both affect the organization for years. Rarely is sufficient money available to accept every project or capital investment proposed. To upper management in many large organizations, both capital investments and projects represent financial and strategic opportunities. The day-to-day management of the project portfolio management is the same as the management of any other group of sizable organizational assets. Both disciplines continually work to optimize their portfolios through the selection and elimination of assets or projects, maximize organizational revenues, minimize portfolio risk, and improve global and strategic goal achieving ability.

Project portfolio management differs from financial portfolio management in that the overall project focus is to use *speed, efficiency,* and *effectiveness* to attain organizational strategic objectives and optimize strategic asset value of project outcomes while minimizing and diversifying project portfolio risk.

5.5.1. The Project Portfolio Inventory

Typically the first step in gaining control of the project portfolio is to compile an enterprise wide project database. This is accomplished by conducting an inventory of projects underway. Benchmarking participants consider the project inventory an important prerequisite to effective portfolio management.

The inventory usually starts with the largest and highest level and works downward to smaller projects. The level of projects inventoried depends upon the organization, and the project inventory or database can initially include any level of detail. Some organizations give each project a number which corresponds with data tracked by accounting and finance organizations. Others start with a listing or log of projects.

> **Standard:** **The best practices project organization maintains a database of all major projects in the host organization.**

In most organizations the process of conducting the project inventory uncovers numerous problems. At the commencement of conducting the inventory, it usually is already apparent that an organizational wide approach to managing multifunctional projects is missing and that no one seems to have responsibility for all the projects in process. Even so, *every* individual in the benchmarking forum that conducted a project inventory said they were surprised at the numerous anomalies uncovered by the process. All reported finding duplicate projects being conducted in various functional and global divisions as well as projects with significant overlap between project activities and goals. Some projects had been in process for years with no termination in sight. Other projects offered no deliverables to support organizational goals and strategies. In a number of cases, senior management was unaware of all the major projects being performed that could impact the strategic success of the organization.

Projects that spanned multiple functional and global areas and did not fit neatly into the organization chart presented particular problems. Examples of such projects were information systems, software, mandated (e.g., government environmental projects), and cross-functional process and systems improvement projects. In each of these categories, there was often confusion, vagueness, and lack of knowledge about the technology. Ownership of project deliverables, scope changes, outcomes, and failures was unclear. Performance was difficult to measure and to track. Frequently these projects fell outside the accepted project approval and governance process. Even so they represented sizeable amounts of expenditures, could impact organizational efficiency, and had potential to become valuable capital

assets (or liabilities). Sometimes, projects were initiated spontaneously in response to a management fad or technological trend. For
example, when new technology products were introduced, some organizations allocated funds with minimal investigation and justification.

Project budgeting irregularities also became obvious. Some departments had greater budgeting flexibility than others. In one organization, $10 million was approved for projects that "just came up"
between budget periods. In many organizations, each functional or
global department created their own projects even though a specific
project might span several enterprise areas. Often when the project
was funded from an individual department, ownership was implied.
There was a feeling that the project deliverables should emphasize
benefit to the funding functional department more than for others
affected. Measurements of performance between projects in the inventory varied from department to department. Consequently, it was
difficult to compare one project with another.

5.5.2. Project Initiation and Selection

Project initiation and selection include all the activities associated
with generating a new project concept and then ranking its organizational value related to the portfolio of existing and other proposed
projects. From this process the project is selected and initiated (or
rejected). The primary document in the project initiation and selection phase is the project charter.

Project portfolio management is constantly faced with the constraint that there is rarely, if ever, sufficient money and resources to
initiate every project desired. Consequently, some form of selection
process must be implemented. The objective of the process is to improve strategic goal achievement while maximizing the potential return of the portfolio and minimizing portfolio risk. When project management is involved with portfolio management and selection,
attainment of strategic objectives is increased because a pragmatic
decision process ensures that each project serves to optimize the overall portfolio value and strength. In addition, the management of specific projects is more focused because the project team is aware of
the relative importance of each project and how it helps achieve the
company's strategic objectives.

Experience of the benchmarking participants is supported by
considerable research that there is a clear positive relationship between proficiency in analyzing and selecting projects and ultimate

project success. Ensuring a symbiotic fit of the project with the portfolio of projects and the host organization's products, services, skills, and resources is also a key factor.

5.5.2.1. The Project Charter

The project charter is the key document associated with the initiation and selection phase (see the methodology chapter for a detailed description of this document). The primary function of the project charter is to communicate all material information needed to make a decision about whether to accept or reject the project. It is the project focused counterpart to the business plan in organizational strategic management. It is the tool used to evaluate the impact of a single project upon the objective of optimizing the project portfolio. Once the project is accepted, signature approval of the project charter launches the planning phase of the project. The project charter then becomes the foundation or starting point for development of the more detailed project plan.

5.5.2.2. The Project Approval Process

All project organizations participating in the Benchmarking Forums follow a structured approach to the initiation and selection of projects. Most organizations include a formal process to ensure that all proposals for new projects are considered, although there are wide variations in the degree of formality involved. After entry into the evaluation process, required is a project charter or other similar document such as a concept paper, business case, or business plan that contains information necessary to make a decision. The charter is presented to a decisionmaking group or individual for evaluation. From this a judgment is made whether to proceed.

5.5.2.3. The Evaluation Group

Projects that cross functional or span global boundaries are usually approved by a senior level evaluation board. Often it is the same board that provides oversight of the portfolio process. Composition of the evaluation board varies but it normally includes vice-presidents, senior executives, and representatives from the various functional areas, strategic business units, and global divisions. The evaluation board typically exercises one of three options: to pass along to the next project phase, abort, or go back and resolve questions or concerns. Usually presentations are made to the board by the sponsoring organization such as the functional or global units.

Numerous project organizations have developed a methodology that includes a phased life cycle. The phase gate approach ensures that all projects in the portfolio are evaluated using an organization wide measurement standard. The phase gate process maintains a disciplined approach and ensures that the project is properly monitored and reviewed. Programmed reviews are established with go/no go decision points and milestones. The danger of the phase gate approach is that it can become rigid and overly bureaucratic. For example, the most complex approval process was described by a participating governmental agency. Their organization required 40 sign-offs to approve projects. The agency's benchmarking representatives said that the process severely limited the effectiveness of the approval process.

As the project size and level of priority declines, usually the decisionmaking authority passes from the evaluation group to individuals. Several benchmarking organizations define each authority level by establishing financial thresholds. For example, one company has a guideline that projects under $50,000 can be approved by supervisory personnel. General managers can approve projects up to $250,000. Projects with larger budgets require approval from the enterprise group.

Some organizations use an automated approval process. Within predefined guidelines, approval requires one signature. A few global project organizations in the benchmarking group approve most projects electronically. A website is utilized for the initiation of projects and anyone can initiate the process. The web site contains a project charter template and defines the categories of projects that can be submitted using this process. Some types of projects require a preapproval from functional or global management before final approval is given. Contract approvals with electronic signatures are returned electronically to the submitter. Users of the system say it works well. It isn't necessary to wait for a formal meeting of the decisionmakers. For projects subject to the process, benchmarkers report that the project approval time has been reduced from four months down to a week. Users caution that using web sites or company intranets works well for a high level overview of a project but is less successful in communicating the details. For projects where there are numerous or significant questions, a face-to-face presentation and defense of the project is required

Projects that impact one functional, global, or strategic business unit, and fall within predefined budget categories, usually are approved by the units themselves. Basically, if the functional organization has the resources they can approve the project within the framework and guidelines of the organizational strategic plan. The project organization is then charged with managing, monitoring, or providing consulting services as a part of the portfolio. The advantage of having the functional or global area approve their own projects is that it emphasizes that authority, responsibility, and ownership remain with the sponsoring area. It also eliminates any problems or ill feelings that arise when the host organization says no or declines to perform a project. The project organization benefits because funding and other resources have already been provided and the project organization is in the position of giving a yes response to all projects offered them. Even by giving the functional and global units ability to approve their own projects, the project organization can still be subject to severe resource limitations. Often the limitation relates to people and specialists.

5.5.2.4. Project Prioritization

Portfolio management necessitates that projects be prioritized. Resources are limited, risk is always a factor, and there are a myriad of strategic considerations. The end result is that not all projects can be performed by the organization. In many cases, projects that are profitable and compatible with strategic objectives are rejected. It is similar to an individual and potential investments. Almost everyone is aware of good investments. They don't make *all* of those investments because they don't have enough money.

> **Standard:** **The best practices project organization ranks projects from top to bottom to determine priorities.**

The prioritization process consists of ranking projects from top to bottom using any of several different ranking scales. For example, the projects might be ranked according to financial return, lowest risk, strategic considerations or any of several other factors. There are also projects that don't need to be ranked. They are nondiscretionary or mandatory. They must be accepted.

5.5.2.4.1. Constraints. Constraints are always a factor. The most commonly encountered is limited resources. Rarely do organizations have sufficient funds to perform every beneficial project proposed. Other resource constraints can be just as severe. Organizations report that scarce skilled personnel such as project managers and subject matter experts regularly place more stringent limitations on project acceptance than funding limitations. The availability of resources can affect the highest as well as the lowest priority projects. There may be two or more important projects that are considered essential to strategic success of the organization, but there may be insufficient resources to conduct all of them.

Constraints may be arbitrary in nature. For example, many new product development projects work under a "no new technology" constraint. The constraint reduces scope changes and emphasizes reaching the market in the shortest time. Often the no new technology constraint is a decision that must be made by senior management.

Potential organization wide risk events also serve as project ranking constraints. The organization may be subject to fluctuations in the economy, the project might be politically sensitive or there could be environmental trends. To ensure that potential major risk events receive consideration, the project charter template encourages that project risk relative to other projects in the portfolio is evaluated. For broad based risks that could impact numerous projects, many evaluation boards engage in "what if scenario" planning. They investigate each of the major risk elements and determine a response. Often the suggested response will impact the projects currently being approved. For example, in a company selling consumer products, the success of a new product could be highly dependent upon the strength of the economy. Even though the economy might be currently strong, the evaluation group might take a more conservative approach by factoring in the potential for a downturn.

A review of constraints is worthwhile for the evaluation board or individual. Spotlighted are bottleneck areas that lengthen the project schedule and increase costs. Constraints associated with mandatory and high priority projects sometimes place an increasingly severe burden on other projects. The *type* of resource shortage can also be a consideration. For example, there may be cash or funding shortages, but the organization has a surplus of people or even excess manufacturing capabilities. This or a similar scenario requires adjustment to the selection process.

5.5.2.4.2. Benefits. By having the project organization pre-rank projects before being presented to the approval committee, benefits accrue. First, the meetings become more efficient and decisions are made faster. Second, it tends to equalize the differences in rigor between functional areas and strategic business units.

Prioritizing projects also provides guidance to management about the degree of emphasis and promotion to be placed on each project. It ensures that the most important projects receive resources and management support and get completed. For example, one large European company reported that they had 400 projects on their strategic "to do" list. After spending $450 million none of the projects was completed. A similar experience was reported by a U.S. chip manufacturer. They listed approximately 100 projects that were considered strategically important. None was completed at the end of 12 months. Both companies initiated a prioritization system. The chip manufacturer was typical. They focused on the "top three" projects. When one was completed, another was added to the list. Several other benchmarking enterprises have adopted the approach.

5.5.2.4.3. Senior Level Guidance   Guidance in the ranking process is required from senior management since the prioritization process should be a direct reflection of organizational strategy. For example, the project group needs to know whether the organization is seeking fast returns on investments or whether it is taking a longer term view. There is a necessity to understand enterprise wide strategic initiatives and the weighting placed on each. Some organizations utilize gap analysis to determine where they are compared to where they should be.

5.5.2.4.4. The Process.   When ranking projects most organizations list the nondiscretionary or mandatory projects at the top. These are projects that must be done. Examples are projects to comply with government regulations or the premillennium efforts to make computers compliant with year 2000 issues. Next ranked are discretionary projects. These are projects where a decision is necessary to determine whether to proceed. Some organizations start by categorizing the prospective projects into those that generate cost savings and those that increase revenue. Strategic considerations and senior level guidance determine which category is ranked above the other. For example, if the organization is in an environment with a rapidly expanding market, emphasis could be placed on revenue growth. If the company is experiencing pressure on profits, it might focus more on cutting costs.

Benchmarkers say that picking the best and the worst is the easiest part of the process. Usually the best projects clearly stand out above all others. The same is true of the poorest. They are so clearly inferior that a great deal of analysis is not required. The detailed ranking analysis applies to the projects in the middle of the evaluation spectrum. These require in depth financial, risk, and strategic analyses to result in an acceptable ranking.

5.5.2.4.5. Pop-Ups. Projects that suddenly appear are labeled by benchmarkers as pop-ups. They usually are a response to an urgent company need and the result of unforeseen circumstances. They could be projects needed to meet rapidly developing competitive situations, regulatory changes, technology changes, and market opportunities. Sometimes they are "break-fix" situations for unexpected warranty or service problems. For pop-ups the tendency is to bypass the detailed analysis requirements of the portfolio ranking process. Benchmarkers would urge that as much rigor be applied as possible within the constraints and pressures of the situation.

5.5.2.4.6. Short Term Focus. Benchmarkers also relate that one phenomenon that tends to occur when ranking projects is the tendency for the portfolio to become increasingly near-term focused and projects to be smaller in size. Reasons given are that the near-term and smaller projects tend to have higher net present values and internal percentage rates of return and faster payback periods. However, they may not be as important in terms of improving long-term organizational goal achievement. To correct the tendency, some benchmarking organizations mandate the review and consideration of long-term opportunities. Emphasized are strategic considerations. One organization in the forum arbitrarily applies 90% of available funding to tactical and reserves 10% for long-term strategic projects.

5.5.2.4.7. Ranking Responsibility. Benchmarkers say the ranking process is somewhat arbitrary and that it is difficult to please all stakeholders. Each is looking at the portfolio ranking from a personal viewpoint. For projects at the top and in the middle there are relatively few disputes. It is the projects at the bottom of the ranking that are subject to termination that arouse stakeholder passions.

One solution is to have each of the strategic business units rank its own projects. The project organization in concert with representatives from the enterprise units can meld the ranked listings with those from other units to form the master priority list. Although more time consuming, benchmarkers say that the process works well and has eliminated many disputes regarding where a project should be ranked.

5.5.2.4.8. Ranking Metrics. Results of surveys of benchmarking participants disclose that 85% of participants in the Benchmark Forum use a formal ranking system for prioritizing projects. Most organizations rank according to strategic factors, financial return, and risk. Benchmarkers also describe other ways of prioritizing projects. Survey respondents disclosed that 76% also ranked projects according to type and 71% ranked according to size. Other ranking method examples described by forum participants are as follows:

- **Stewardship.** The concept of stewardship overshadows the ranking process in several organizations in the Benchmarking Forum. Organizations where stewardship is an important consideration include those entrusted with other's funds, those engaged in the manufacture of dangerous products, and those producing life saving products and services. Stewardship is an important consideration in organizations that are heavily regulated, have safety oriented organizational values, and are working toward being environmentally sensitive.
- **Customer impact.** Benchmarking Forum organizations that rank by *customer impact* determine how the project maximizes the customer experience. Ranking by client type, targeted client segments, and project impact on end user needs are also conducted by some benchmarking organizations.
- **Resource availability.** Resources are sometimes a key ranking factor. For example, projects might be ranked by schedule availability or when the project can be performed. If there is restricted availability of funds, projects could be ranked according to timing of budgeted cash flows.
- **Political.** Benchmarkers say that sometimes politics and cultural factors are important ranking considerations. One attendee said that after they conduct their thorough and detailed analysis, the chief executive officer makes the final project selection by taking into consideration the various political wild cards.
- **Miscellaneous.** Other ranking methods described include the degree of probability of success and the availability of resource capacity such as manufacturing facilities.

Standard: **The best practices project organization selects projects by evaluating and ranking strategic factors, financial benefits, risk, and other appropriate factors.**

As mentioned, the most popular ranking methods are by strategic factors, financial return, and risk. A detailed discussion of each follows:

5.5.2.5. Strategic and Subjective Factors

Organizations that are proponents of the strategic approach to management first evaluate and determine the project's strategic fit with the host organization's strategic plan. The project is evaluated to determine how it interfaces and complements the critical success factors for the organization. By default, priority projects are those associated with the highest level of enterprise strategies.

Benchmarkers stress that often the subjective strategic elements are more important to the decisionmaking process than the quantitative or mathematical ranking methods. They also say that the process is conceptually more difficult than simply drawing a line and accepting projects above the positive net present value point. For example, evaluation of strategic factors means that sometimes the highest ranked project according to financial return could be rejected, e.g., the project could be in a totally unrelated line of business. Inversely, other projects might be accepted even though their financial return is low and risk is high, e.g., the project might be needed to keep competition at bay or it could be mandated by the government.

Researchers have disclosed numerous strategic factors that have positive correlation with project success. Rothwell found that understanding and responding to user needs was important. Benchmarking survey respondents validate the importance of determining the fit of each project with the strategy of the organization and existing services and product lines. Synergies, complementary linkages, and ways that projects can benefit and support each other become apparent. Projects can be selected that enhance organizational competencies and strengths and improve weaknesses. By emphasizing strategic value ranking, spotlighted are projects that have minimal strategic connection but are consuming resources.

When investigating the strategic aspects of the project portfolio, benchmarkers relate that there is need to constantly focus on the future. One benchmarking organization said that they periodically predict what the portfolio will look like five years in the future, what projects will have been completed and the impact on the organization. This information is then compared to the existing strategic plan to determine if there is a match.

5.5.2.6. Financial Return

After evaluating strategic importance, the traditional financial approach to ranking alternatives is by potential financial return. One benchmarker suggested, "Just look for the big piles of money!" Financial analysts would agree. Project organizations tend to rank all investment opportunities intuitively by how much money the project is expected to earn compared to its cost. Techniques used commonly involve the calculation of net present values (NPV), internal rates of return (IRR), and payback periods. From these calculations, the projects can be ranked according to those generating the most revenue down to those generating the least.

5.5.6.2.6.1. Present Value, NPV, and IRR. To evaluate a portfolio of projects necessitates ranking them according to expected financial return. Problems arise because rarely do any two projects have the same timing of cash flows. The three project opportunities shown in Table 5.1 serve as a good example. All three projects require a $1,000 investment. In return, project A receives relatively small amounts of return in the early years but expects a large payoff later in its life. In actual dollars, it forecasts $1,700, the largest amount of any of the three prospects. Project B has sizeable funds coming early in the project life, but its total return at $1,500 is the least of any option. Finally, Project C forecasts a steady flow of funds over its life and records the second best lifetime return in actual dollars at $1,625. From the standpoint of financial return, Project A would appear to offer the greatest potential.

Intuitively, most business people express uneasiness with the timing of returns associated with project A. The returns are minimal

TABLE 5.1 Example Projected Returns for Three Projects

Year	Project A	Project B	Project C
0 (investment)	($1000)	($1000)	($1000)
1	300	1,500	525
2	300	250	525
3	300	250	525
4	300	250	525
5	1,500	250	525
Total	$1,700	$1,500	$1,625
Benefit-cost ratio	2.7	2.5	2.62

until year five at which point a sizeable payoff is forecasted. Many things can go wrong in five years and risk is higher than for project A where the payoff is faster. To level the field of evaluation present value techniques are used. Present value calculations have become easy as a result of numerous inexpensive financial calculators that eliminate the math. Applying the math to the alternatives above results in the following data assuming a minimum of 10% return is required.

Ranking method	Project A	Project B	Project C
NPV	$882.34	$1084.06	$990.16
IRR	20%	79%	44%
Payback	3 years 4 months	8 months	2 years

Adjusting all the cash flows to current values results in a different view of the project alternatives. Now Project B would appear to be the most financially favorable project. Its net present value is highest at $1,084. 06. Net present value is interpreted as the profit the project will earn *over and above* its 10% minimum rate of return. The internal rate of return for Project B is a whopping 79% per year. Project A now becomes the least favorable project with the lowest net present value at $882.34 and lowest internal rate of return at 20%.

5.5.2.6.2 Payback. The payback method measures the period of time it takes to recover the initial investment of the project. In the preceding case, payback for project A would be approximately three years and four months; for project B it would be eight months, and for project C approximately two years. Another example would be that if a truck were required for moving dirt on the project its cost might be $50,000 and it is projected to save $5,000 per month. The payback would be 10 months.

Until the advent of computers the payback method was the most popular method used to evaluate the financial return of investments. It is simple to calculate and easy to understand. It reflects conventional investment wisdom because it encourages fast returns. Its disadvantages compared to present value calculations are that no cash flows are considered after the initial payback period and it fails to consider the time value of money.

5.5.2.6.3. Benefit-Cost Ratio. A way of comparing multiple projects with varying levels of cash investment and return is to calculate total benefits expected to be received against total costs fore-casted to be spent. In its most simple version, the total cash coming in (revenue) is divided by total money to be spent (the costs). The result is a ratio. In the example above, the benefit-cost ratio of Project A is 2.7, of Project B is 2.5, and of Project C is 2.62. Project A has the greatest amount of revenue ($2,700) compared to the $1,000 invest-ment. The ratio of 2.7 for Project A indicates that 2.7 times as much revenue is expected as funds invested. Based solely on this ratio, it could be concluded that Project A would represent the highest return of the three projects.

Note that in this example the costs of each project are the same ($1,000) so the ratios calculated reflect the same conclusion that would be reached by simply observing the total dollar return from each project. In practice there could be dozens of projects being evalu-ated with each project having different costs and different cash flows over its life. The benefit-cost ratio provides a way to compare and rank the projects.

The advantage of the benefit-cost ratio is that it is simple and easy to understand. Most people intuitively understand that the project should bring in more revenue than it costs. Hence the benefit-cost ratio should always be over 1.0 before a project would be ac-cepted. Projects with ratios under 1.0 are predicting that there will be higher costs than revenue. These are money losing ventures and would normally be avoided. Of course sometimes projects with bene-fit-cost ratios under 1.0 *are* accepted. Examples are mandated proj-ects such as those required by law, strategic projects, e.g., developing a product to fill a gap in the project line, and warranty and service projects.

The disadvantage of the benefit-cost ratio is that it ignores the time value of money and the timing of cash flows. Money expected three years in the future is valued the same as cash received immedi-ately.

5.5.2.7 Portfolio Risk

It could be tempting to simply select the projects with the highest return. Unfortunately, high financial return projects are often sad-dled with corresponding high rates of risk. Consequently, the supe-rior project manager evaluates the risk elements associated with each project. These are then compared with the expected financial returns. Risk analysis is more difficult than the process of financially ranking projects because there are few well accepted, simple, and easy to un-derstand project risk ranking tools.

> **Standard:** **The best practices project organization evaluates individual project risk related to the risk of the overall portfolio.**

Project management involves two types of risk analysis. The first occurs during the project selection process and involves the overall risk inherent in one project when compared to another. The second application of risk analysis occurs during the project planning process. It involves a detailed risk analysis of a specific project.

Project portfolio risk analysis is much more difficult than the process of estimating and ranking projects according to their amount of financial returns. Various methods of ranking portfolio risk are now discussed.

5.5.2.7.1. High-Medium-Low. The most common and easily understood method of risk ranking consists of estimating whether the project is high risk, medium risk, or low risk. Projects can then be grouped into categories.

5.5.2.7.2. Weighted Risk Factors. Various key project portfolio risk elements can be listed and given a numerical value. For example, the projects might require new, unproven technology, capital investment could be large and stretch the company's resource base, or there could be stringent delivery time constraints. Each of these factors can be listed on a grid:

	Unproven technology	Investment high	Delivery stringent	Total
Project A	5	5	5	15
Project B	4	4	5	13
Project C	3	3	3	9

Note: 5 is highest risk, 1 is lowest risk.

In this example, Project C would have the lowest number of risk points. This information would be compared with the amount of potential financial return as well as the subjective factors in making the decision whether to select the project.

5.5.5.7.3. Company Project Portfolio Beta. The mathematical average risk for all the projects in the organization's portfolio is labeled the company project portfolio beta. The beta is used as a baseline to evaluate the risk of new projects and to diversify the risk of the overall portfolio. For example, if a company has the majority of its project investments focused on the domestic market, it could choose to diversify and reduce overall portfolio risk by focusing project investments on the foreign market.

5.5.2.7.4. Probability. Although most organizations evaluate project risk by estimating high-medium-low, it is also possible to apply probabilities of success to various alternatives and subalternatives. An example in the world of project management would be to develop a probability decision tree with each branch of the tree showing the expected probability of occurrence and the expenses or gains associated with the occurrence. The method is particularly appropriate for complex projects with numerous options unfolding at various project phases.

5.5.2.7.5. Simulation. An easy way to remember simulation is "change one thing at a time." Typically simulation analysis refers to the use of spreadsheets such as Lotus and Excel. For example, an income statement or cash flow might be developed for a new product. If the impact of a pending recession were to be evaluated the team could estimate the impact, (say, a 20% reduction in sales), plug that into the spreadsheet, and then observe and evaluate the results.

5.5.2.8. Combined Financial Return/Risk Ranking Tool

Benchmarking organizations report the use of various tools that simplify the process of comparing complex projects with a multitude of financial and risk factors. A calculation can be performed to simplify the ranking process. Expected sales, profit, and risk are compared with cost in the following equation:

$$\frac{\text{Sales} \times \text{Profit per sale} \times \text{\% Probability of success}}{\text{Cost}} =$$
$$\text{Ratio of return vs. cost}$$

If a company expected to sell 500 units of a product with an expected profit of \$250 per unit, with an estimated 80% probability of success and a cost of \$50,000, using the formula above, the ratio of return versus cost would be calculated as follows:

$$\frac{500 \times \$250 \times .80}{\$50,000} = 2$$

The interpretation of the ratio of 2 in the calculation is that the project is expected to bring in twice as much *risk adjusted revenue* as cost expended. The formula does not account for the time value of money and other subjective factors.

5.5.2.9. Combining All Major Factors

Some decisionmaking tools simplify the complex scenarios in the selection process.

5.5.2.9.1. Combined Weighted Ranking Matrix. A numerical ranking can be performed using the same technique as discussed in the risk section. The investigators rank the portfolio of projects according to financial return, risk, and any subjective factors felt appropriate. If some factors are felt to be more important than others they can be weighted accordingly. For example, in Table 5.2, a maximum of five points each can be applied to financial return and the lowest amount of risk. Only a maximum of three points each are given to each of the subjective factors.

5.5.2.9.2. Grids. The graphic presentation of the relationship of projects with each other is effective when there are a few factors that dominate the selection process. As demonstrated in Figure 5.1 this particular selection team judges that market and product variables are most important. Each project is plotted and can then be visually compared. As many grids can be used as felt necessary to depict all important selection variables. Once a pattern is determined, several grids can be combined into one summary grid that has several factors on each scale. The grids are less effective when there are numerous projects and many factors that impact success.

5.5.2.9.3. Pairwise Comparison. When projects or alternatives being compared are broad based, nebulous, and have numerous dissimilar characteristics, it can be difficult to rank and compare them as a group. Often in such cases, it is easier and simpler to compare each individual project with each other project. Thus the vari-

TABLE 5.2 Project Ranking Matrix

	Expected financial return	Risk (low)	Subject factors			Total points
			Product benefits	Market growth	Strategic fit	
Project A	5	4	3	3	2	17
Project B	4	4	2	1	3	14
Project C	4	3	1	3	1	12

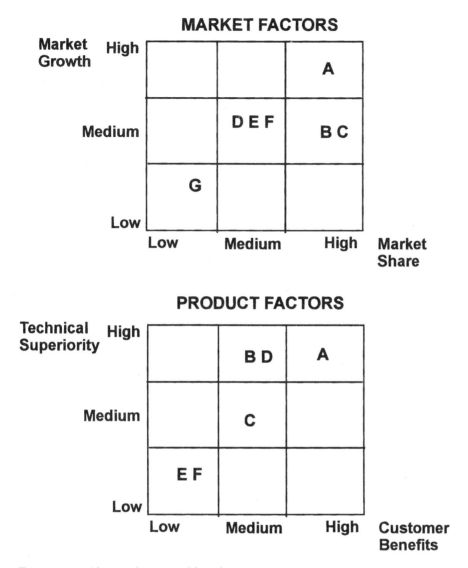

FIGURE 5.1 Key project considerations.

ables are limited. The pairwise comparison chart provides a ranking of the resultant conclusions. For example, in Figure 5.2 project A is preferred overall to project B. Therefore a 1 is placed in the A row and a 0 is placed in the B row. Next, project A is less favorable than project C so a 0 is placed in the A row and a 1 is placed in the C row. The process continues until all the projects have been compared individually with each other.

	A	B	C	D	TOTAL
A	X	1	0	1	2
B	0	X	0	1	1
C	1	1	X	1	3
D	0	0	0	X	0

FIGURE 5.2   Pairwise comparisons.

5.5.2.10. Portfolio Analysis and the Need for Relevant Information

The information available to assist in making the selection decision can range from almost nonexistent to nearly infinite. Consequently, the seasoned participant in the selection process will be selective in analyzing only relevant information.

5.5.2.10.1. Minimize Sunk Cost as a Factor.   Sunk cost is represented by money already spent or invested in a project. In project as well as financial management sunk cost is not considered when evaluating future project investments or additional investment in a continuing project. The objective of the approach is simple. It is to protect against the human tendency to "pour good money after bad." It is particularly appropriate in project evaluations where scope has changed numerous times and it is difficult to evaluate future returns related to past targets and investments. In those cases, the evaluators ignore sunk cost and evaluate the potential future returns related to forecasted costs.

5.5.2.10.2. Include Opportunity Cost.   When evaluating project alternatives, opportunity costs are pertinent considerations. They represent opportunities sacrificed by making one project investment related to another. For example, if funds are limited the selection of one project means that another must be sacrificed. It is valid to evaluate the impact of the opportunity lost when comparing more than one project.

5.5.2.10.3. Think Incrementally. Often the selection of projects becomes confusing. For example, it is common in high technology projects for a new product to cannibalize or take away from the sales of existing products. In those situations the recommended approach is to evaluate the incremental sales expected from the project selection.

References: Clark and Fujimoto, 1991; Dwyer and Mellor, 1991; Gupta and Wilemon, 1990;   Daft et al., 1988; Cooper and Kleinschmidt, 1987; Maidique and Zirger, 1985; Lindsay and Rue, 1980; Cooper, 1980; Tung, 1979. For general guidance about portfolio management review the capital budgeting chapter from *Contemporary Financial Management* (1995) by Moyer, McGuigan, and Kretlow.

6

Portfolio Management of Project Planning

6.1. PORTFOLIO PROJECT PLANNING

After approval of the project charter by the portfolio group or individual, the planning phase commences. The planners expand the project charter into the detailed project plan. The project plan includes the work breakdown structure, project schedule, budget, risk analysis, functional and technical specifications, communications plan, quality standards, and any other details felt necessary to execute the project (see methodology chapter for details). At completion, the plan will include approval signatures from the customer or sponsor and sometimes major stakeholder groups as well.

The primary objective of the portfolio governance process during the planning phase is to ensure that the project plan remains faithful to the original project charter and host organization's strategic plan. The project plan details the *tactics* that will be executed in support of the organizational strategy and should contain an appropriate level of information to ensure project success. Although scope changes may occur in the project during the planning phase, they should be reviewed to ensure they maintain focus on organizational goals.

Benchmarkers emphasize that the key individual in successfully developing the project plan is the project manager, not the portfolio manager. Multiple research efforts conclude that the project leader is the best positioned individual to convert the project charter into a workable team strategy and the project plan is the tool that defines the strategy.

> **Standard:** **The best practices project organization monitors and supports the project plan process to ensure that the plan satisfies the strategic objectives set forth in the charter.**

The involvement of portfolio management in the planning process is one of coordination, support and direction. The amount of participation depends upon the number of projects in the portfolio and the structure of the project organization. As a minimum, the portfolio manager will monitor progress toward milestones set for expected project plan completion dates. Portfolio management involved in reviewing specific project plans will usually utilize a checklist of items from a project plan template such as found in the methodology chapter. Benchmarking Forum attendees report that there are specific project planning subject areas that often call for extra scrutiny. These are now discussed.

6.1.1. Project Plan Level of Detail

The advice of benchmarking participants is that the project plan should include a level of detail *appropriate* for the project's complexity and risk. If there is a question about the level of detail required, the generally accepted heurism is to *plan more rather than less.* The logic is that most problems that occur during execution of the project can be traced to insufficient planning. Extensive planning also means that extra care has been taken to identify all factors that could influence the project. From a cultural or political view, planning also provides an aura of professionalism. Particularly on projects that are predictable and have been performed many times, such as building a house, detailed planning has a direct positive correlation with accuracy in attaining schedule and budget objectives.

> **Standard:** **The best practices project organization encourages an *appropriate* level of detail in project plans.**

At the same time, benchmarkers caution portfolio managers that the planning process is expensive and time consuming. Every excess day spent on the project plan is a day the execution of the project could have commenced. Benchmarkers remind that the operational purpose of the project plan is to make the execution of the project proceed faster, more efficiently, and more effectively. In some cases, excessive planning is counterproductive and hinders attainment of project goals. The research of Brown and Eisenhardt showed that on some high technology projects *the more planning that was conducted, the less successful the project outcome*. Projects where excessive planning harmed the project team's efforts were those where the technology was changing while the project was in progress and the customer demanded that the latest technology be incorporated. The same logic applies to planning in situations where the external environment could be expected to change rapidly and dramatically.

Excessive planning can also result in project plan rigidity. As a result the team loses flexibility in reacting to changing events and taking advantage of unexpected opportunities. According to another research study, too much planning and analysis can result in poorer quality decisions. The researcher found that analysis and planning of the smallest detail sometimes induces the planner to lose sight of the big picture and get sidetracked on insignificant issues.

6.1.2. Project Scope

A working definition of project scope is that it includes all aspects of the project that involve expenditure of money and resources when a change is made. The scope of the project is first outlined in the project charter. After the charter is approved the planning process begins. As stakeholders review and critique the plan, they may uncover areas that require revision. As a result, scope changes are likely to occur during the planning phase and often are major. They could represent modifications to basic functional and performance characteristics, amount and type of resources to be used, schedules and milestones, and even deliverables.

> **Standard:** **The best practices project organization requires definition of project scope including all aspects that involve additional resources if changed.**

The view of benchmarkers regarding scope changes made during the planning phase is that if they must be made, *now* is the best time. Benchmarking participants stress that changes made during the planning phase are easier to incorporate in the project plan than if the same changes are made after project execution has commenced. Once the fundamental decisions have been made and the project is underway, it becomes progressively more difficult to incorporate major scope changes. At later stages of project execution, all the project elements, activities, and deliverables are so interrelated that even a small change can affect multiple aspects of the project. Eventually the point is reached where there is almost nothing that could be described as a small change.

Even though early scope changes are encouraged, they still must be analyzed to determine their impact on the project plan as it reflects the host organization's strategy. The process always includes project owners or sponsors, stakeholders, and others whose signatures are needed to signify approval of the project plan.

References: Toney 1996; Brown et al., 1995; Clark and Fujimoto, 1991; Makridakis, 1990; Lamb, 1987; Miles and Cameron, 1982; Snow and Hrebiniak, 1980.

6.1.3. The Work Breakdown Structure

A core oversight activity of portfolio management is to ensure that every project has a work breakdown structure. It is the preferred tool used by portfolio management to view and understand the overall nature of the project. By presenting the project in a simplified, graphical format, project complexity is reduced and nebulous goals are transformed into a more easily understood picture. The work breakdown structure is an ideal focal point during discussions about scope changes, project tactics, alternatives, and opportunities.

> **Standard:** **The best practices project organization requires a graphical depiction of the work breakdown structure that shows an appropriate detail of deliverables and activities.**

Benchmarking portfolio managers report that project managers and project team members sometimes assume that senior managers and other stakeholders have a high degree of knowledge about the project. This is often not the case. The work breakdown structure solves the problem by refreshing memories about the deliverables and overall project scope. It reflects the common understanding between the sponsor, project and host organizations, and other stakeholders about the nature of the project.

Benchmarkers also judge that portfolio management *oversight* is necessary because there is a temptation for project managers to skip the development of the work breakdown structure. The lure is to "save time" and proceed directly to the project scheduling software and begin entering activities. When this "easier and faster" approach is taken, the individual ends up with numerous pages of activities and lines depicting the Gantt chart. When asked to describe the project, the only way it can be shown or discussed is to present the multipage listing of the Gantt chart. Portfolio managers agree that a better way is to show the project in the form of a work breakdown structure.

References: Pfeffer, 1992; Zangwill, 1992; Maideque and Zirger, 1990; Gupta and Wilemon, 1990. Eisenhardt, 1989; Cooper and Kleinschmidt, 1987; Feldman and March, 1981b.

6.1.4. Clear Performance Measurement Metrics

Benchmarkers avow that a key aspect of the planning process is defining the manner in which project performance will be monitored and tracked. In a sense, the development of the project plan and associated measurement metrics represent an application of management by objectives. The project team works with the project sponsor or owner and host organization to develop mutually agreed upon targets whose attainment is ideally gauged by clear, observable, measurable, and achievable performance metrics.

Measuring the performance of projects and the results of project management are areas that differentiate the professional project organization from others. Surveys of the benchmark participating organizations conclude that 92% formally measure performance of projects. Schedule variance is measured by 81% of participants, quality of the end product by 69%, budget variance by 58%, application of project manager skills by 42%, lead time to market by 31%, and customer satisfaction by 20%.

Emphasis on performance measurement is not found to the same degree among the general population of organizations. The research of Lynda Radosevich determined that only about one-half of the 28 companies she studied use *any* form of metrics to evaluate project performance. Of those that do, most collect them infrequently, with about 20% gathering metrics monthly and 8% looking at them once or twice per year. About 10% of metrics users collected and evaluated the measurements weekly.

On information systems projects executed for internal organizational customers, research conducted by Dr. Bill Ibbs found that "very few" employed formal performance measurement programs. Since most information systems projects studied were conducted within host organizations, profit and other empirical measures of performance were usually not the motivating goal.

Benchmarkers agree that critical success factors should be measured throughout the project life cycle. To do so requires the development of a disciplined approach to monitoring and reporting project performance. Included are portfolio wide measurements that provide baselines and average values against which individual project performance can be evaluated. The detail of the process ranges from scheduling periodic reviews of overall performance to the monitoring of daily performance measurements.

6.1.4.1. Metrics Characteristics

Effective portfolio management requires measurement standards that are simple, easy to understand, interpret, and compile. Metrics should lend themselves to being rolled up into portfolio wide statistics as well as being comparable to similar information from other projects in the portfolio as well as industry wide data provided by benchmarking organizations. Benchmarkers say that the best metrics encourage behavior that results in positive project and portfolio results. They reflect corporate strategy and the needs of the customer and are compatible with organizational culture. Best practices metrics identify root causes of problems and point to possible solutions. They inspire continuous improvement.

> **Standard:** **The best practices project organization requires clear, easy to apply and understand metrics against which project performance and progress are measured.**

Superior metrics are inexpensive to collect, interpret, and retain. They are simple to calculate and gather, easy to understand, and intuitively logical. Their measurements lend themselves to a graphical presentation and are straightforward to communicate and interpret. The most successful measurement approaches utilize commonly accepted concepts and principles.

Best practices metrics are reliable and provide a verifiable measure of current project and portfolio status. They are valid measures of what they claim to measure and are supported by the results of other metrics. They are independent and insensitive to outside variables and influences. They are timely and available when the information is needed.

As an example of the application of the performance measurement process, the *Best Practices Report* of September, 2000, described the metrics developed by First USA Bank. The organization categorized their metrics as follows: Performance metrics measured the organization's delivery of products and services compared to the customer defined needs for quality, timeliness, and competitive costs. Stability factors measured the development and improvement of new products and improved services as a reflection of the strategic plan. Compliance metrics ensured that project management best practices were followed. Capability metrics determined whether customer needs and requirements were satisfied and if business needs of the organization were attained. Finally, improvement metrics targeted the overall performance of the project management process.

6.1.4.2. Metrics Problems

Benchmarkers warn that the measurement, compiling, analyzing and reporting of performance metrics are expensive. The methodology adds time and cost to individual projects as well as the management of the portfolio. A resource consuming exercise associated with the portfolio tracking systems is keeping the information updated. There is empirical support for the benchmarker's views. In Radosevich's research, organizations using metrics reported that the activity represented 5 to 8% of an information system's operating costs.

At the other end of the bell curve, benchmarkers testify that there is a tendency for some project organizations to focus so intensely on measurement metrics and tools that the face-to-face management of projects is sacrificed. The measurement process often is an intrusion into the productivity of the project team. Further, as any student would verify, testing of performance and scrutinizing of the results by others is not a popular activity. However, as the students' *teachers* would respond, measurement and communication of performance increases the learning rate and accentuates individual performance. The same is true of project related performance metrics. Their objective is to fuel improvement and to verify that the project is remaining on track.

Another problem reported by benchmarkers is that sometimes the project team places too much emphasis on easy-to-obtain quantitative metrics such as lines of code. More important are the metrics that define satisfaction of customer and organizational needs and goals. Often these metrics tend to be subjective, nebulous, and difficult to define, *but are still necessary*. The client or stakeholder may find it challenging to specifically identify and articulate their needs related to the project. Sometimes on vague and rapidly changing high technology and information systems projects, customers are unclear about the capabilities of the technology, the final deliverables, and their role as project owner. For these projects, performance measurement clarity and preciseness is more challenging and will probably be more subjective in nature.

6.1.4.3. Metrics Used

Examples of metrics employed by benchmarking organizations are outlined in the following subsections.

6.1.4.3.1. Customer Satisfaction. There is general agreement that the ultimate measure of individual project performance is customer satisfaction. One benchmarker remarked that the key to measuring project performance is as easy as answering the question, "Is the customer happy?" It is not always a straightforward process. Benchmarkers say that expectations of customers and their view of project performance vary widely depending upon their perception of how the project was handled. A poorly run project to one customer is well run to another. Often the customer's view of the project is enhanced by the application of good people skills and communication by the project manager and project team. Benchmarkers relate examples of projects that were considered less than optimum by project team members but still received good ratings from the customer. The reality is that customer satisfaction is as much related to the quality of the relationship with the project manager as it is the application of "technically" correct project execution.

> **Standard:** **Best practices project organizations measure project performance by evaluating customer satisfaction.**

Factors that the customer and other stakeholders consider critical to project success should be identified and agreed upon early in the relationship. Usually this is accomplished during the original discussions with the project customer about needs and project deliverables. Ideally these will have been defined and detailed in the project charter. Development of measurement metrics during the planning phase should place priority on determining progress toward meeting the customer's stated needs and objectives.

> **Standard:** **Best practices project organizations define critical success factors in measuring project goal achievement.**

Best practices project organizations have a formal customer satisfaction measurement process. One benchmarking organization asks customers to rate the project every two months. Others request and review ratings quarterly. One benchmarking organization conducts an annual survey of customer satisfaction that covers all projects. As the project is being executed and milestones are reached, additional customer input can be solicited.

In addition to the formal rating process, some benchmarking organizations have representatives from the project management group meet with the customer and talk about each area of performance. They judge that the survey instrument analysis is less important than the process of meeting and talking. Another benchmarking firm felt that personal contact with the customer was so important that they scheduled quarterly "spontaneous drop-by" visits or personal telephone calls by regional vice-presidents.

6.1.4.3.2. Milestones. For tracking the overall performance of the project, the *preferred* metric by benchmarking organizations is the milestone. For measurement purposes, the term "milestone" is often used interchangeably with phase gates and go/no go decision points. Milestones are specific dates or times scheduled for completion of activities, deliverables or project phases. They are convenient points that encourage evaluation of the project and evaluation of its performance before proceeding to the next step and incurring additional expenses. Upon arriving at the milestone, it is a simple matter to observe whether the project is ahead or behind schedule. For example, assume that a new home is scheduled for completion on December 1st. If December 5th has been reached and the house is not complete, it is intuitively logical that it is four days late.

Standard: **The best practices project organization encourages the use of numerous milestones to serve as goals and against which to measure performance.**

Most project organizations report project status in achieving key milestones. Some show the information in the form of a master calendar that shows each project and key milestones. Others track the "milestone hit rate" by measuring the percentage of key activities or deliverables completed on time. They say the advantage is that it keeps stakeholders and the project teams focused on deliverables rather than day-to-day metrics and technical issues. Users of milestone reporting say it is particularly appropriate for monitoring virtual project teams in broadly dispersed global locations. They enable the portfolio manager to see at a glance the status of each team.

Milestones are popular because they are easy to use and understand. They encourage focus on schedule control and meeting deadlines. They serve a dual purpose of providing clear goals for the project team while providing portfolio managers precise points in time to evaluate and review overall project performance. Milestones can be set for sub projects as well as the overall project.

Benchmarkers report that frequent milestones accelerate project speed because they force the project team to focus on short-term minigoals. As a result, stakeholders review the project more often. When projects are off course, corrections can be made earlier in the project life cycle. The project organization can motivate and synchronize team energies by rewarding project teams for achieving milestones.

6.1.4.3.3. Deliverables. Some best practices benchmarking organizations improve project goal achievement by tying deliverables to milestones. Proponents of the concept say that a key to improved customer satisfaction is being able to "see" the progress of the project. The practice has been successful in information systems and software organizations. One best practices software development group has a guideline that *"every project should deliver something of value that the customer can see every six months."* Benchmarkers applying the philosophy say that on such projects, tying deliverables to milestones is a demanding exercise because it necessitates that project managers convert intangibles into periodic visual or physical deliverables.

A side benefit of planning the project as a series of deliverables tied to milestones is that the process encourages modularization or breaking the large project into subprojects. For example, one benchmark organization modularized a 5,000-line software project into 26 small projects. Proponents of the modularization and deliverables concept say that project performance becomes easier to measure as does the management of the project. Risk is mitigated because problems can often be contained within one module. Some groups have expanded the concept by having the functional elements of the project teams break the major subtasks into deliverables as well.

6.1.4.3.4. Critical Path Monitoring. The schedule status can be evaluated by providing "critical path status" that shows the number of days delay for the critical path.

6.1.4.3.5. Periodic Reviews. For projects where it is difficult to establish milestones or phase gates or tie reviews to deliverables, the project can be reviewed periodically, e.g., every six months.

6.1.4.3.6. Non–Time Based Performance Measurements. The performance on many projects is difficult to measure using time based milestones, phase gates, and go/no go decision points. An alternative measurement metric is necessitated on projects where performance measurement is required *between* milestones. For example, the deliverable on a long-duration government project could be a single prototype aircraft with a construction period spanning three years and the only project wide milestone being the delivery of the prototype. Even on smaller projects, the project status may need to be reviewed between milestones.

Milestone based performance measurement also has limited suitability for complex projects with multifunctional teams and where each functional area is working at its own pace. The result of such an approach is that each functional area arrives at the end of project phases at different points in time. Many speed based projects encourage the functional areas to proceed immediately into the next phase. Some groups overlap the *entire phase* by having each new phase commence before the preceding phase is complete. A similar situation is found on large projects with numerous sub projects where each subproject is also proceeding at its own pace.

The common characteristic of all of these projects is that the overall schedule performance status of the project is difficult to anchor with a single time based measurement such as a milestone. Consequently measurement metrics based on other standards such as funds expended and work packages complete are used. Examples are earned value where dollars budgeted at any point in time are compared with the budgeted cost of work actually performed. A more simple and easy-to-understand method than earned value is the percentage complete calculation.

- **Percentage complete.** Benchmarking participants report the use of two types of percentage complete calculations. Both are simple to calculate and easy to understand. One is to compare the number of *work packages completed* with the total amount planned for the project. The other is to compare *funds expended* with those scheduled to be spent for the entire project. For example, assume the project is the construction of a new product prototype. Scheduled time to the prototype completion milestone is six months. The number of work packages in the total project is 600 and budgeted funds are $1,200,000. If a status report is requested at the end of month three (half way through the project), it is a simple matter to count work packages complete and add up the money spent. In this example, if 300 work packages are complete (50% of the total) and $600,000 has been spent (50% of funds budgeted), the project is on schedule.

  The most simple and easy to understand percentage complete concept described by benchmarking participants is the "checkbook budget." The total amount budgeted for the project is entered as the opening balance. As the project expends funds, deductions are made from the project checkbook. It is a simple and easy to understand approach that helps the project manager visualize the relationship between each expenditure and the budget at completion. The exercise communicates to project managers that once they pass the zero balance level, they are in an overdraft, or negative territory.

It also is a simple matter to compare funds spent with the degree of project completion.

- **Earned value.** A somewhat more complicated form of percentage complete calculation is "earned value". Earned value consists of two basic components. The first is a set of formulas used to evaluate *schedule* variance. The second group of formulas determines *cost* performance.

Earned value is primarily used by the U.S. government and associated contractors. A survey of 60 participants in the Top 500 Benchmarking Forum concludes that only a handful use earned value for *schedule* management. The major complaint is that the schedule measurement component of earned value calculations is complicated and difficult to understand because it uses money to measure time. Also, experience of participants has been that when earned value is used, it is necessary to spend excessive amounts of time educating employees in its use and interpretation.

For schedule performance, earned value compares the budgeted cost of work performed to date (BCWP) with the budgeted cost of work scheduled to date (BCWS). The formula is BCWP − BCWS to obtain a dollar amount, or BCWP/BCWS to provide an index or percentage. It is different conceptually from the more simple percentage complete formula because it works only with budgeted amounts rather than actual amounts.

The costing portion of earned value calculations reflects the conventional cost accounting approach and is widely used and accepted within for-profit organizations. The costing portion of earned value compares *estimated* total work package costs with total *actual* work package costs. No distinction is made between number of resources used or unit costs for each resource. The formula to determine earned value costing variance is budgeted cost of work performed *minus* actual cost of work performed (CV = BCWP − ACWP) to find a dollar amount, and budgeted cost of work performed *divided* by actual cost of work performed (CPI = BCWP/ACWP) to determine a percentage or index of cost variance.

The simplified nature of the earned value cost calculation lends itself to estimating completed costs and efficiencies required to meet targets. The estimate of the cost of the completed project assuming that current cost overruns will continue is determined by dividing the cost index by the original total budgeted amount for the project (EAC = CPI/BAC). The degree of efficiency needed to bring the project in on the original cost target is determined with the formula TCPI = (BAC − BCWP)/(BAC − ACWP).

6.1.4.3.7. Financial Measurements. The most frequently applied performance measures for projects are financial. Their use is well established in management. They are reasonably easy to use, apply, and understand. The variances provided are precise. Various examples of financial measures follow:

- **Profit Impact.** In for-profit organizations the ultimate measure of performance is profitability. Although strategic objectives are of utmost importance, the reality is that many senior executives of corporations are driven by profits. When evaluating project performance the constant overshadowing question is, "What's the bottom line?" The ultimate measure of project success is the impact of each individual project's contributions and impact on organizational profits.
- **Budget and cost performance.** The major resource with which portfolio management is normally charged is money. Once a commitment is made during the initiation and selection phase to allocate funds to the project, portfolio management administers and monitors the budgeting and expenditure of those funds.

  Although it initially would seem a simple matter to budget funds for the project over its entire life, benchmark respondents report that in practice the process is complex. The reason is that accounting systems in large functional organizations are seldom compatible with funds flows and reporting associated with large multifunctional projects. Budgeting, cost tracking, and reporting systems in functional organizations typically are structured around *time* periods, whereas budgeting for projects focuses on *groups of activities* and project phases. For example, budgeting for a three-year project encompasses the entire project time span. The incompatibility arises because the project will likely be executed in an organization with an entrenched *yearly* budget process.

  Once the project is underway, additional anomalies are encountered. Most accounting systems are geared to be "closed out" every year. Inefficiencies and consumption of scarce management time occurs when functionally oriented accounting departments try to force the project cost and variance reporting process into the constraints of yearly or other periodic financial reviews. More problems are encountered when entering individual *project* expenses since many accounting systems are structured around *departments* and department wide expenses. Problems also result when an attempt is made to roll-up expenses for the entire project portfolio into top level totals for evaluation comparisons. All of these troubles are compounded when working with projects where the technology is evolving and the budgets for future phases are vague, difficult to define, and subject to dramatic change.

As a result of these quandaries, few of the benchmarking organizations operating in functionally oriented organizations report a smooth interface between project and functional organization budgeting and cost reporting. Exceptions are companies that traditionally conduct large projects as a core activity of the organization (e.g., defense contractors and pharmaceutical companies). Companies that place all project budgeting under the wings of the portfolio management group also report success. The benefit is that an independent accounting system can be developed, administered, and integrated with the master organization budget and accounting reporting processes.

Aside from the difficulties of meshing accounting systems, most benchmarking groups use conventional budgeting and costing metrics from the world of accounting during the planning process. Their application is as simple as estimating or budgeting the cost of each work package. Once the project is underway, the amount budgeted can be compared with the amount actually spent. The result is the variance. There are two types of variances commonly used: the *amount* of any resource used (e.g., the number of people) and the *unit cost* of each resource (e.g., the estimated cost per hour of a person).

- **Activity based budgeting and costing.** Most large projects lend themselves to activity based costing. The cost of an entire activity is measured over the period that the activity is in progress. It is particularly appropriate if a project spans multiple accounting periods. Traditional accounting is geared toward closing the books at the end of major periods such as year end. With activity based costing, the costs and revenues continue to be recorded over the life of the project.

   Activity based costing can improve project team motivation. At Safety-Kleen, the managers in the factory were given bonuses based upon how they performed as measured by a yearly budget. Implementation of an activity based accounting system made it apparent that a better reward system was one that encouraged longer-term focus. In particular, increased emphasis was given to reducing overall unit costs of production and materials processed. To measure the type improvements rewarded, necessitated was an accounting system that recorded the financial aspects of activities spanning multiple accounting periods.

- **Life cycle budgeting and costing.** The entire life of a *product* is the measurement standard for life cycle costing. It is often encountered in the project management environment, particularly in association with new product evaluations. Life cycle costing differs from other costing methods because it generally includes revenue as well as material and labor costs. When evaluating the life cycle of capital goods, machinery, and many consumer products, there are sizeable cash flows after the sale of the product. Included are such items as repair and service components, maintenance, potential to move up to larger models, added features, trade in potential, and salvage value. Numerous products have greater cash flows *after* the product has been sold than before.

  Life cycle costing can have dramatic impact on new product selection and strategy. For example, some computer printers are sold at low prices with little if any profit made from the product sale. The marketing strategy is to generate lifetime profit from the sale of toner.

- **Job order budgeting and costing.** Job order costing applies costing concepts to jobs that have a specific beginning and ending and include clearly definable units of production. In a project environment it could pertain to the activities or segments between each milestone, individual work packages, or even the total project.

  Job order costing measures the cost to produce a group of products or services. For example a "batch" in a sporting goods factory might be 50 baseball bats. Total material and labor costs to build the batch of bats would be recorded. To determine the cost of one bat would entail dividing the total cost recorded by 50.

- **Process budgeting and costing.** The difference between job order and process costing is that job order costing measures the cost of a group of *units of production* and process costing measures units produced in *a specific period of time*. Process costing is used in operations such as mining and petroleum production. Some observers judge that process costing has little application to projects. However, if a project has few milestones, over a period of time it begins to take on the attributes of a process flow operation. Further, if one considers earned value calculations, it will be noted that the calculations apply more to projects with continuous flows rather than those with many small, segmented milestones. In a way, the calculations associated with earned value could be considered an attempt to measure progress in a process flow environment.

- **Standard costs and variance analysis.** During the project planning process, the costs for each of the tasks comprising project work packages are estimated. These estimated costs become the "standard" costs of the project. They have counterparts in nearly every industry. In the auto service business they are termed flat rates. The primary reason they are used is because it is easier to use an estimate than to calculate actual rates, and actual costs are not known. Once the project is underway, actual costs begin to be accumulated. Comparing the actual costs with the estimated or standard costs results in variances.

6.1.4.3.8. Other Quantitative Measurements. Financial metrics reflect only one aspect of project performance. There are numerous other quantitative measures that give an expanded view of activities.

- **Personnel utilization.** Personnel are the major resource used on many projects. Numbers of people in each pay and skill category can be estimated during the planning stage. During the execution phase, people usage can be compared with the estimates. Variance from the original project plan can be measured.
- **Measure forecasting accuracy.** Several best practices organizations evaluate projects on the basis of accuracy in meeting goals. For example, a project might be under *or over* schedule by, say, 5%. Either a positive or negative result would be equally accurate and receive the same management response.
- **Nonproject time.** Benchmarkers report that often in large organizations there are numerous distractions that limit time spent performing project related work. Some maintain "nonproject" time cards. On the card are entered such items as non-project related meetings.

6.1.4.3.9. Subjective Performance Measures. Many of
the best measures of project performance are subjective in nature.
Previously discussed was the topic of the customer's overall opinion
about project success. Following are other subjective measures used
by the benchmarkers to judge project performance.

- **Strategic success.** One of the most important subjective
  measures is to judge whether the project has achieved its tactical goals in support of the organization's strategy.
- **Observation of results.** The objective of many internal systems projects is to generate improved organizational efficiencies and improved performance. Benchmarkers say that one
  method of measuring the degree of success of the project is to
  observe whether promised improvements have resulted. For
  example, an improved efficiency is often accompanied by staff
  reductions. Observation can determine if this has, in fact,
  happened.
- **Stakeholder input.** The opinion of stakeholders about the
  degree of project success is an important measure. The experience of benchmark participants is that project success varies
  with time and often depends upon the view of the stakeholder.
  Some best practices project organizations query stakeholders
  for input in the same manner as the project customer is solicited. In particular, benchmarkers say that it is wise to assess
  satisfaction at different levels in the organization.
- **Team member opinion.** One of the most valuable subjective
  measures of project performance is to query the project manager and team members about their general feelings about
  the project's progress and degree of success. Benchmarkers
  say the process works well but requires skillful questioning
  and insight to paint a complete picture. One benchmark participant said that he concentrates on the, "Yes, buts." For example, the influence of team spirit will inspire some project
  teams to declare themselves successful. By asking the team
  how they could do things better, they are encouraged to talk
  about some of the problems encountered. The benchmarkers
  also are mindful that even the most successful projects have a
  multitude of problems. To determine whether problems being
  described are significant or unusual requires that they be considered in the context of problems encountered on other similar projects.
- **Measure value of quality.** Benchmarkers avow that measuring cost of quality is a noble goal, but is difficult to do. As

a minimum, quality specifications can be monitored for conformance. A few best practices project organizations have made anecdotal estimates of the overall value of project quality. One group formally queries customers about the degree of satisfaction with project quality. Most benchmark attendees agree that the overall topic of quality and quality standards and specifications should receive attention and be measured to the best of the group's ability.

- **Open issues.** Review questions that are within the project team's control and have not yet been fully resolved.
- **Project team attitude.** Evaluate the general disposition of the project team. Query regarding people and human resource issues.

6.1.4.4. Portfolio Measurements

All of the measurement methods used by a project organization can be aggregated over a period of time and numerous projects. The aggregation of measurements gives the project organization capability to evaluate improvement and set goals for future projects. Some participating organizations roll up the aggregates of all their projects for the year. They show the results of each specific project as well as all projects in total. Status and performance measurements are graphically portrayed for all stakeholders to evaluate. A baseline of performance in any category of performance can be developed.

> **Standard:** **Best practices project organizations aggregate project goal achievement and/or benefits of projects.**

Once the baselines have been set and performance measures established, numerous benefits result. It is now possible to measure the impact of problem identification and corrective action for individual as well as groups of projects. A continuous improvement plan can be implemented and results measured. For example, specific areas of project performance can be targeted (e.g., improving schedule forecasting accuracy). Motivational programs such as giving bonuses and awards for improved performance can be implemented.

Marginal projects are more easily identified and culled earlier in the development process. One project management group reports that they identify and terminate projects on the average every three months. At the inception of their performance measurement efforts, it took approximately one year to terminate a clearly losing project. Other organizations report that they measure the time to identify problems and the number of project problems corrected.

Portfolio management lends itself to measuring lead time to market since the measurement is obtained after termination of the project. It is an important measure of performance in organizations producing new products. One large telecommunications company in the forum conducts a "momentum analysis" to measure the benefits of reducing lead time to market for new products. The analysis consists of establishing a benchmark of measurement and then comparing later results and efforts against this effort. This is done by examining several projects and determining when the projects began and when they ended. By emphasizing lead time to market, the telecommunications company judged that they increased sales and reduced costs by approximately $3-billion.

6.1.4.5. Measure the Cost of Professional Project Managers

It is the role of the project organization to show the value added benefits, and that professional project managers will improve project performance well in excess of their cost. This issue is of particular importance because the professional project manager is generally more expensive than the nonprofessional, and sometimes potential customers view professional project management as an optional service. Consequently, best practices project organizations measure the cost of professional project management. The cost is then compared with project output. Several project groups have a line item "cost for project management" or "construction management." They measure the percentage of the project cost that is management related. Several project management groups quoted rates from 1.5 to 5% of total project cost as allocated to project management.

6.1.4.6. Cost of the Project Organization

Portfolio management makes it possible to evaluate the benefits resulting from implementation of the project organization. Usually the measurement is more subjective than analytical. Targeted is improved strategic focus, increased financial return, and reduced overall portfolio risk compared to the historical approach to managing individual projects.

6.1.4.7. Performance Measurement for Changing Projects

A final unanswered question is how project performance is measured when numerous scope changes have been made, such as when scope changes have been encouraged to mask project performance. The accepted methodology is to use the sunk cost approach. Specifically, all previous expenditures and changes are ignored. The project is evaluated based on expected future benefits compared to the additional cost and investment required.

References: Pennypacker, 2000; Schnapper, 2000; *Best Practices Report* 2000; Radosevich, 1999; Ingram, 1998; Ibbs and Kwak, 1997; Ness and Cucuzza, 1995.

6.1.5. Overoptimistic Targets

A problem for the portfolio management process is the tendency for project planners to *grossly* underestimate the amount of time required to execute the project and the amount of funds to be expended. Pennypacker's *Best Practices Report* of August, 2000, describes research that shows that nearly one-half of projects are over cost approximately 100 to 400%. The research further concludes that 86% of projects are late in meeting deliverables.

For functionally organized enterprises, the seeming inability of project groups to accurately estimate time and cash flows is a frustrating concern. Most *functional* units operate under conditions where they *consistently* budget cost and resources almost to the dollar and individual person. Host organizations rely heavily on an accurate and predictable budget and planning process to schedule funds allocation and cash flows as well as to manage other resources such as personnel. For professionals nurtured in this environment, it is anathema to witness and even participate in the shoot-from-the-hip approach taken by many project budget and scheduling exercises and then to witness the wild inaccurateness of the actual project results.

Benchmarkers say that there are numerous reasons why project groups underestimate cost and schedule targets. The most common is that planners forecast times and budgets without considering the impact of potential problems and risk. Each work package has a time and cost prediction that is based on the assumption that all will go well. Practitioners say this is almost *never* true in real life. Innumerable problems arise, risks are encountered, and unforeseen uncertainties present themselves.

There is ample evidence that the inaccuracies can be overcome. Several benchmarking organizations dependably estimate within plus or minus 5% of the actual project cost and schedule attained. Organizations that bid information systems and software projects to external clients reliably meet cost and schedule projections while maintaining profitability. One benchmarking organization successfully bids and executes 7,500 external projects per year. The project managers know the cost and schedule accuracy of their project on a daily basis. Even some best practices project groups who execute internal information systems projects are accurate in their predictions of time and budget targets.

Best practices project planners compensate for overoptimism in planning by adding contingency funds to the budget, factoring in problems and risk in the planning equation, and adding a queuing buffer to the project schedule.

> **Standard:** **Best practices project organizations use contingency funds and queuing buffers to ensure schedule and budget accuracy and to factor in risk and uncertainties.**

A budget contingency fund compensates for risk and unknowns. The size of the fund is based upon the degree of risk and uncertainty associated with the project. Benchmarkers note that sometimes planners attempt to compensate for risk by "padding" each work package with extra funding. For performance measurement purposes, the preferred method is to *be as accurate as possible* in work package forecasting and to offset risk with a separate contingency fund.

For scheduling risk and unknowns, one solution is to utilize software or planning tools that consider the impact of pessimism. For example, most scheduling software packages have capability to consider optimistic, most likely and pessimistic time estimates for each work package. The formula associated with PERT charts is Optimistic time + 4 times the Most likely time + Pessimistic time, all divided by 6 [(O + 4ML + P)/6]. For example, say that a person typically requires 60 minutes to drive to work (the most likely time). If the individual were putting together a schedule and the task was "driving to work," the 60-minute time would probably be the amount entered.

The logic of the PERT calculation is that the 60-minute time is overly optimistic and fails to display a range of performance. There are a multitude of problems and events that can cause sizeable delays. A more realistic estimate would consider the impact of inclement weather, accidents, and other delays. When asked for the most pessimistic or the longest time to drive to work, usually the answer will be in the range of 120 minutes or so.

When asked to state the fastest (the optimistic) time to get to work, the individual would probably indicate the time to be about 50 minutes. The response mirrors the reality of project life that seldom is the optimistic time all that much faster than the most likely time. Most people are working about as hard as they can work and people drive about as fast as they can safely proceed.

By plugging the most likely, pessimistic, and optimistic drive times into the formula, the following result is obtained: [50 + 4(60) + 120]/6 = 68 minutes. The resulting "expected time" of 68 minutes is a considerably longer time estimate (13% longer in fact) than the original most likely time of 50 minutes. If the same logic is applied to an entire project planned without considering the impact of risk and uncertainties, it could be assumed that total project time would be underestimated by 13%.

The formula also provides an expected range of performance. A simplified standard deviation (range) can be calculated by using the formula: (pessimistic—optimistic)/6 [(P − O/]6). In this case the standard deviation would be 11.7 minutes. From this, the prediction could be made that 68% of the time (one standard deviation) the expected time for the individual to drive to work is 68 minutes, plus or minus 11.7 minutes.

Although the formula factors *expected* risk into the schedule, remaining is the issue of *unexpected* uncertainties and unknowns. For example, projects could encounter technology improvements, the maneuvering of competition, and dramatic economic changes. As in the case of the project budget, a contingency is necessitated. The schedule contingency is labeled the *queuing buffer*. It provides extra time or a cushion to compensate for unknowns. The queuing buffer is usually tacked on to the end of the project (i.e., at the end of the queue). For example, if preparing a presentation for senior management, a few days might be added to provide flexibility in putting together the presentation. Ideally the presentation would be ready a couple of days before the presentation date. If problems are encountered the extra days are utilized. The end result is that the project comes in before the deadline and, one hopes, with time to spare.

Benchmarkers address the psychology of making time predictions and then being late. They say that it is better to forecast that the project will take longer than expected rather than less (i.e., include the queuing buffer in the estimate). One example quoted is the psychology of people waiting to be seated in a restaurant. If the best estimate of actual seating time is 15 minutes, the waiting customers would be told that the wait could be 20 minutes (includes a queuing buffer). If seating occurs at the 15-minute milestone or even a few minutes late (into the queuing buffer) but still sooner than 20 minutes, people are generally happy because they are ahead of schedule. The same logic applies to projects where on-time presentation of the deliverables is consistently rated by customers as the most important factor impacting their sense of satisfaction.

Standard: **The best practices project management organization is consistently on time and on budget.**

Benchmarkers stress that the project organization is a key element in encouraging cost and schedule accuracy. It can create an "accuracy" state of mind through schedule and budget performance evaluation. The project organization serves as the role model in communicating that the world of project management is dominated by time, including starting meetings and keeping appointments precisely as scheduled, completing work packages when due, and remembering that achieving time targets is generally considered the most important measure of performance by project customers.

6.1.6. Signatures

When the project plan is complete, signatures are required to authorize the execution phase of the project. Generally the signatures of the customer or project sponsor are included. Often, particularly for internal projects, the signatures of other stakeholder groups are included as well.

The signature gathering process is an important element of the planning phase. For portfolio management it ensures that stakeholders understand the project and are willing to formally commit to the plan. People who are required to sign a document often take greater care in scrutinizing its contents. The scrutiny process serves to find areas needing improvement and change and reduces later scope changes. The end result is greater overall understanding and commitment to the plan.

References: Eisenhardt and Tabrizi, 1990; Gersick, 1988.

7

Global Project Portfolio Strategy: Execution, Control, and Termination

Once the existing projects throughout the enterprise have been inventoried and ranked, new projects selected and initiated, and their charters expanded into detailed project plans, emphasis transfers to the day-to-day strategic execution and control of the portfolio. As in the previous phases, the overshadowing objective is to achieve the strategic goals of the host organization. The project organization relies upon metrics and tools to monitor the progress of each individual project related to the overall portfolio. Conflict is guarded against because the individual project managers' focus has shifted from development of tools (i.e., the charter and project plan) to the management of the people. Maintaining project scope related to the original plan is of particular importance.

In carrying out the strategic initiatives of the host organization, an important success factor for portfolio management is the ability to optimize performance of projects dispersed geographically. Benchmarkers would attest that a most important success component is to select superior project managers. Creating a virtual project management office optimizes communications. Scope change is an important strategic success factor and requires formal approaches and constant vigilance. Status of the projects in the portfolio is monitored through the use of software and traditional measurement tools.

7.1. PROJECT MANAGEMENT EMPHASIS SHIFTS FROM TOOLS TO PEOPLE

When moving from the initiation and planning processes to the execution phase, the *project manager's* focus shifts from *tools development* to *people management*. During the initiation and selection phase, emphasis was on the preparation and presentation of the project charter and the subsequent approval process. Throughout the planning phase, attention was directed toward developing the project plan and gaining scope endorsement from stakeholders. Now, in the execution and control phase, concentration by the project manager shifts to the people management of the project. The tools become secondary to face-to-face and personal communications as primary management aspects of the project.

Benchmarkers note that the shift in emphasis is an important consideration for portfolio managers. At the same time the project manager has transferred from tools focus to people management, portfolio management will continue to concentrate on tools; e.g., formal scope change control, variance reports, and milestone reviews. To prevent conflict, benchmarkers say it is important for portfolio management to remember that their continuing emphasis on tools can be an intrusion into the effective people management of the project. Consequently, a guideline given by benchmarkers is to request sufficient information to ensure that the portfolio is implementing the organization's strategy but not to demand so much that project management productivity is hindered.

Standard: **The best practices project organization requests the minimum amount of reporting needed from the projects to effectively monitor portfolio performance.**

7.2. SELECT SUPERIOR PROJECT MANAGERS

Much of the knowledge required to govern and manage dispersed projects is well entrenched in history. Management pioneer Henri

Fayol was chief executive officer of a French mining company with over 10,000 globally dispersed employees in the 1850s. The Catholic Church has effectively directed a large global organization for well over a thousand years. The Roman Empire controlled a geographical area that encompassed most of Europe and North Africa from about 300 BC to approximately 500 AD. Later, the British Empire spanned the globe with colonies. With the advance of modern communications technology, the administration of globally dispersed projects has become easier.

> **Standard:** **The best practices project organization places highest priority on selecting superior project managers.**

In all of the historical examples of managing remote enterprises, there was one consistent, dominating leadership best practice; that was to carefully select and train the project manager or emissary being sent to the remote environment. Clearly understood was that communications would be limited or even nonexistent, and responsibility for success was being placed on the individual project manager. Hence, an associated foundation stone was the training and preparation of the project manager. In addition, benchmarkers and the evidence of history agree that for remote project teams and management to work, the project manager or remote leader should be given the authority required to maximize strategic performance and ensure timely completion of the project deliverables.

7.3. CREATE A VIRTUAL PROJECT OFFICE ENVIRONMENT FOR GEOGRAPHICALLY DISPERSED PROJECT PORTFOLIOS

In modern times, a key to the maximization of the probability of success is for the best practices project organization to duplicate the benefits of working in close proximity. In effect, it seeks to create a *virtual project office environment*. The primary interaction method is through electronic communications.

> **Standard:** The best practices project organiza-
> tion creates a virtual project office
> environment that duplicates the
> benefits of collocation.

7.3.1. The Problem

Under ideal conditions project teams would work in close physical
proximity with each other. Even with modern communications technology there is consensus among practitioners that distance *normally*
reduces project speed, efficiency, and effectiveness. Researcher Allen
agrees. He measured distance between desks and team members in
engineering groups and found a direct relationship with the positive
communication of scientific and technical knowledge. Other researchers found that communications effectiveness declined when workers
were more than 20 feet apart. Zangwill's research found that communication was maximized at little effort when the entire team was located in one room.

When managing a geographically diverse project portfolio, close
proximity is rarely possible. Teams are located in different buildings,
facilities, departments, divisions, strategic business units, and global
locations. One benchmarking organization reports that they have
projects being executed in approximately 1,500 locations worldwide.
When managing broadly dispersed project portfolios it should be recognized that they will be *inherently less efficient* than projects located
within one arena. For these situations, it is important to consider the
impact of physical proximity of teams on project speed, efficiency, and
effectiveness.

Leadership and strategic coordination of the geographically dispersed array of teams face greater challenges than encountered when
exercising the traditional approach of "managing by walking around."
In most cases there is no visual confirmation of the completion of project activities. It isn't even possible to see if people are actually working. Participants are no longer privy to the informal communications
and gossip that convey the context of events taking place. Physical
interaction through verbal and nonverbal signals and facial expres-
sions are all lost when using most remote communications tools.

There are also reduced efficiencies, cultural clashes, isolation,
and trust concerns. Maintenance of central files and resource materials becomes critical. The technical support must be problem free, easily accessible from all over the world, and available 24 hours a day

and seven days a week. Costs of travel and communications are higher. The setup and maintenance of remote offices and facilities adds expense. For example, the research of Cascio estimated that a mobile or home office for a single employee is $3,000 to $5,000 with yearly maintenance running $1,000.

7.3.2. The Virtual Project Office

The inherent problems associated with the geographical dispersion of projects in the portfolio can be solved, or at least partially solved, by creating the virtual project office. In some ways, the modern management of the project portfolio is well suited to a virtual environment. Historically, client based projects such as encountered in sales, marketing, engineering, consulting, and construction have tended to be located remotely—often at the client's worksite.

7.3.3. Communications

The major technological improvement stimulating virtual project office success is improved communications. Remotely located project teams now have access to much of the same information, support, and resources as groups located in the central office. Communications advancements such as telecommunications, email, and the internet, have improved the ability to monitor and direct the strategic effectiveness of broadly dispersed projects, project teams, and individual team members. In certain cases, research is recording productivity *increases* resulting from the use of globally dispersed teams compared to being collocated.

The content of communications with global teams is important. Researcher Cascio describes a study of 29 global teams that communicated predominately through email. The highest performing teams tended to begin their messages with social information. They greeted each other as friends, introduced themselves, and gave personal background information. Benchmarkers gave examples of project teams that compensate for remoteness by duplicating social events on-line. They meet on-line to have lunch and schedule after-work social hours, birthday parties, weddings, and baby showers. Others post photos of themselves so people can attach a face to a name.

Cascio's research found that the top performers established clear roles for each of the communicating individuals. They maintained a positive attitude, enthusiasm and eagerness and were oriented toward project action items. Successful written communications included numerous verbal rewards and recognition for accomplishments. He also found that even one pessimistic individual would undermine the communications and effectiveness of a group of team members. Communications include regular electronic meetings. Generally, written communications include more questions with broader distribution of copies. An objective is to keep all associated stakeholders in tune with the various conversations taking place. Regular updates and status reports keep everyone informed of progress. A constant assessment of progress of each team toward milestones and meeting deliverables keeps all teams focused on their ultimate objective.

Portfolio managers can take steps to ensure that interaction is expanded beyond the limitations of email, web pages, and chat rooms. Telephone conversations, video conferencing, and physical visits to the projects work well.

7.3.4. Training Project Leaders to Work in Virtual Environments

Benchmarkers advise that training should include knowledge and skills needed by project leaders working in dispersed geographic environments. How to communicate effectively in the electronic medium is important because nearly all interaction is based on written dialogue. Instruction covers methods of providing feedback and criticism and exchanging ideas in a positive, nonthreatening, and non adversarial manner. Social protocol and socialization behavior encompasses disclosing appropriate personal information, expressing appreciation for efforts and ideas, when and how to apologize and accept apologies, rephrasing of unclear communications, and acknowledgement of assignments. The importance of manners and the application of common cultural values are also explored. Examples are methods of communicating emotion, acknowledging receipt of messages and the necessary degree of promptness in responding to questions. Also emphasized is the use of specific software communication tools and software.

7.3.5. Benefits

Benchmarking participants report exceptions to the research showing productivity decreases associated with distance. Some global organizations are finding that the virtual project office environment is conducive to improved project speed, efficiency, and effectiveness in specific situations. One trend resulting from the improved communications environment is that workers are no longer tied to an office or central working location. Companies specializing in software projects have achieved savings from the change. Cascio's research of virtual workplaces found numerous examples. IBM reported savings of 40 to 60% from the elimination of offices. Included was the cost to install high speed data lines and computers in the homes of remote workers. Northern Telecom said that they save $2,000 per person per year in real estate costs alone. Productively increases were recorded as well with IBM reporting gains of 15 to 40% and US West recording as high as 40% increases. Much of the gain was the result of saving commuting time to and from work. Profit improvements also were recorded. Hewlit-Packard doubled sales per salesperson by letting them locate in remote locations. Anderson Consulting found that consultants without permanent offices spent 25% more time with their clients.

On projects that can be conducted simultaneously around the globe, production increases are resulting from *global scheduling*. At least two benchmarking participants report dramatic output improvements from "following the sun." For example, a U.S. group codes software on the project during their normal eight-hour shift. At the end of the day, the data are electronically transferred to the group's counterparts in Hong Kong. The Hong Kong team continues the coding work over their day shift. The project is then handed off to the European operation. After the European group has made their contribution the project is once again relayed to the United States where the process repeats itself. By having the work follow the sun, in theory production could be tripled. Benchmarkers caution that for follow-the-sun scheduling to be effective, the project should be clearly defined and structured.

References: Cascio, 2000; Top 500 Benchmarking Forum minutes, 2000; Snow et al., 1999; Knoll and Jarvenpaa, 1998; Zangwill, 1992; Keller, 1986; Keller and Holland, 1983; Bourgeois, 1980; Allen, 1977.

7.4. MANAGING SCOPE CHANGES OF PROJECTS WITHIN THE PORTFOLIO

The project organization is an important element in maximizing performance of a global or multi-functional project portfolio. The project organization is usually the only group that is charged with ensuring that *all* projects in the portfolio remain focused on attaining the overall host organization's goal. Unless the project organization monitors the projects and ensures that they are remaining on target and scope, it is easy for the projects to become diverted. As a result, their tactical value could be diminished.

As introduced during the discussion about the project planning phase, scope changes are alterations to the project that affect the schedule, budget, functional and technical specifications, and any other aspect of project execution that involves the expenditure of funds or resources. The emphasis on monitoring and controlling the strategic facets of scope change is continued with even greater consideration during the execution phase.

> **Standard:** **The best practices project organization has a formal process to manage project scope changes.**

Involvement of the project organization is vital to ensure that projects continue to reflect their approved plans and support the organization's strategy. Although the project organization may not personally approve scope changes, it creates the procedures and provides oversight to see that the appropriate stakeholder groups scrutinize scope changes. It is an important activity. Zangwill's research determined that the most successful project groups apply discipline and formal methodology to control scope changes. A speed based approach is maintained by minimizing time consuming distractions and diversions resulting from uncontrolled scope changes. Scope change control ensures that alterations are promptly approved that add value or increase revenue. At the same time, questionable modifications can be evaluated by an independent group in a more arm's length and pragmatic manner.

In general, scope changes create havoc with the structured approach to project and portfolio management. The experience of benchmarkers is that projects with shifting scopes *will* fail to meet the original time and cost targets. Almost every change of scope alters the financial management baseline, budgeted cost of work performed, budgeted cost at completion, functional and technical specifications, and schedule, including milestone and review points, deliverables, and project goals. Scope changes make it more difficult to measure performance, monitor progress related to the original project plan, and compare current progress with prior efforts and milestones.

On projects with rapidly changing technology the impact of scope changes can be even more intrusive. As changes are made to incorporate new technology and then to correct its associated problems, projects become more nebulous, ever changing, and harder to define. Research discloses that often the new technology is incompatible with existing technology, is not fully developed, does not deliver its promised improvements, is unreliable, and is of poor quality.

Benchmarkers also caution that there is a governance problem that necessitates the involvement of the project organization. Often, it is the *project manager* who generates and even encourages, scope creep and project change. In most cases, the project manager is attempting to be cooperative and to maximize customer satisfaction. In other cases, the scope changes may be more self-serving in nature. Ingram's research identified numerous examples where project managers allowed uncontrolled scope changes to make it more difficult to measure performance, disguise failure, and justify continuing projects that might otherwise be candidates for termination.

Project organization oversight is also necessary because almost all other participants in the project process are inclined to encourage rather than discourage scope changes. On high technology projects where the project is changing while the project is underway, the customer often desires the latest technology. There are numerous other stakeholders that place constant pressure on the project manager to make project changes. In the view of the requestor of the change, nearly all scope change requests and recommendations are made for good and positive reasons. They may add value, adjust the project to changing conditions, be legally mandated, or serve as the medium to incorporate improvements and correct oversights. They provide for functional and technical updates and correction of errors. Sometimes external events or changing conditions require the entire project direction to change.

When executing external projects, scope changes represent opportunity to generate incremental revenue and profit. Scope change is a key component of the sales strategy in many organizations. The companies bid projects at low prices to obtain the contract with the intention of making the profit from the predictable scope changes.

7.4.1. Scope Change Responsibility

A cornerstone of the scope change process is to make sure that the responsibility for scope changes is clearly defined and understood by all participants. Benchmarkers agree that ultimate responsibility for scope change rests with the project owner or sponsor. It is similar to a married couple building a house. Any changes that are made during the building process will eventually be billed or charged to the paying couple. Benchmarkers advise that it is the building contractor's (project manager's) responsibility to evaluate the changes and make sure that the homeowners understand the impact on the project schedule and cost.

Benchmarkers also suggest that training should communicate to the project manager their role as the "front-line" component in the scope management process. The project manager and team are charged with screening, communicating the general impact, managing, and controlling most scope changes at the project work site. For scope changes that pass through this initial screening process, the project organization activates the formal scope change methodology. The scope change procedures make certain that a formal review and control process is applied. By communicating that the project manager and team have initial responsibility for scrutinizing scope change requests and eliminating as many as possible, prevented is the danger that the project organization becomes the scope "police force." The project manager based screening process also serves to minimize the expensive and time consuming application of the formal scope change methodology.

7.4.2.   Changes Subject to Scope Change Control

Benchmarking participants report different approaches to deciding the type changes that require formal approval. Most allow incidental changes that have no material impact on the project to be made on the spot. Others require formal board approval of changes that impact cost over a certain percentage, such as 10%. Some project groups require customer signatures on even the smallest of changes. Particularly in the early stages of the project, they say that it is important to enforce the project change discipline. Advocates of this approach say that stakeholders are essentially trained to know that when they request changes, a formal process is initiated and that the change will probably require more money and schedule changes.

7.4.3.   The Scope Change Process

One thing is certain, there is constant pressure for scope change to occur. To manage them requires a formal process that defines how the proposed changes will be evaluated. The objective of the approach is to insure approval of positive or revenue enhancing scope improvements while preventing unjustified or negative scope changes. The process is always approached in a formalized manner.

The details of the scope change control process are communicated to project customers and other stakeholders during the planning phase. In contractual project relationships, the scope change process should be spelled out in the contract. Described is the paperwork that initiates the change and description of the approval process. The scope change process always involves obtaining customer signatures.

The management of scope change involves analyzing the change to determine its impact on the project plan. Often the process incorporates other stakeholders and team member evaluations. Many organizations have single or multiple control committees composed of members from the various functional groups affected by the project. On large projects authorization may be required by as many as four change control boards. One board might review market impact, another manufacturing implications, another serviceability and another price and cost effects. The boards are charged with maintaining the integrity of the manner in which changes are made. They also make sure that all stakeholders totally comprehend the effects of the change.

It was mentioned that sometimes the project manager allows and even encourages scope change. As scope changes pile up, it becomes increasingly hard to evaluate project performance. For projects such as this, which have become vague and nebulous as a result of numerous scope changes, the sunk cost approach is suggested. Previous scope changes that make the evaluation of future actions confusing and difficult are treated as unrecoverable sunk costs and ignored. The project is then evaluated in reference to the expected future benefits compared to the additional cost and investment required.

References: Ingram, 1998; Zangwill, 1992.

7.5. MONITORING AND CONTROLLING
PORTFOLIO STATUS

Monitoring and control of the project portfolio are judged by benchmarking participants to be important functions of project organizations. When forum participants were surveyed to determine the biggest advantage provided by the project organization, about 20% reported that it was improved portfolio and individual project control combined with consistent enterprise wide reporting. The reason given was that the quality and approach to project control by portfolio management represented potential for sizeable savings and increased strategic effectiveness. As previously mentioned, Ingram found that approximately 70% of the projects he analyzed came in materially late, over budget, or failed to meet the client's expectations. The Standish Group data concluded that 86% of software projects came in late and/or over cost and incurred significant scope changes.

Benchmarkers say that applying effective monitoring and control processes dramatically improves these results. Providing oversight of all projects in the execution phase ensures that individual projects are conforming to their project plans and the portfolio as a whole is maximizing the strategic objectives of the organization.

Standard: **The best practices project organization monitors the project portfolio in sufficient detail that no single project deviates materially from its original project plan without portfolio management being aware of the deviation.**

Benchmarkers report a high degree of success resulting from portfolio monitoring. One reason is that benchmarkers say that they start with the assumption that *every* project will encounter some type of problem. Observing the performance of numerous projects makes it easier to spotlight deviations by individual projects. Scrutinizing measurement data for problems in their early stages helps portfolio managers become aware of developing dilemmas even before they are acknowledged by project managers and project teams. One Benchmarking Forum attendee said, "At 20% completion the raw data shows what some project managers won't admit until they are 80% done."

Benchmarkers say that all prior efforts expended on developing the project charter and the detailed project plan are of little value unless the portfolio of projects remains on track and is executed speedily, efficiently, and effectively. They avow that the success of the project organization depends upon the successful completion of the individual projects in the portfolio and that the ultimate responsibility for executing the project rests on the shoulders of the individual project managers. The role of the project organization is one of oversight and support. Viewed is the progress of individual projects related to the overall portfolio and the original project plan. Progress toward achieving goals and meeting deliverable objectives receives constant attention by portfolio management.

7.5.1. The Monitoring Process

All of the benchmarking participants engaged in portfolio management indicate that they have a formal monitoring and control process.

They utilize progress and problem reports combined with face-to-face meetings with the project managers, periodic conference calls, video conferencing when needed, and communication of the data to senior management and other stakeholders. Interim periodic reports provide measurements against pre-established metrics. To generate the information, each project manager or team representative updates the project's measurement data.

> **Standard: The best practices project organization has a formal monitoring and oversight process for the project portfolio.**

Benchmarkers involved in the portfolio oversight role stress the need to take an enterprise wide perspective. In addition to the needs of the project organization, information is gathered for stakeholders, senior management and other global and organizational units. To supplement the overall portfolio view of the project, variance statistics obtained from the monitoring and control process are usually modified to fit specific organizational and stakeholder needs. For example, they might be presented from a geographic or global view, segregated by functional area or strategic business unit, or categorized by size.

The reporting and information submission periods vary widely depending upon the organization and the quantity and nature of the projects in the portfolio. Some benchmarkers have a daily or weekly status report for all projects. Others require project managers to submit monthly reports against measurable standards.

The information from the periodic reports, meetings, and other communications is generally consolidated into an overview report and submitted to senior management and governing committees. The submission may take any one of several formats. Included are executive briefings and presentations, steering committee reports, periodic progress/problem reports, and face-to-face meetings with individual senior managers to discuss specific projects and problem areas.

When projects deviate from standards, the portfolio managers seek to identify the nature of the problem in an informal manner. If necessary, review meetings and other appropriate actions are scheduled with the project teams. As information is accumulated it is communicated to stakeholders.

7.5.2. Variance Measurement

Nearly all forms of portfolio monitoring and oversight use *variance measurement* methodologies. Specifically, the project plan states what the project (in terms of schedule, budget, or specifications) is *forecasted to do*. Once the project is underway, results determine what the project is *actually doing*. The difference between the two is the variance. The variance can apply to any of the many performance measurement metrics defined during the planning phase.

> **Standard:** **The best practices project organization uses variance analysis as a tool to monitor and control deviations from the project plan.**

Benchmarkers stress that the report template for the rolled up information should be simple and easy to understand. Several benchmarking organizations use a red-yellow-green signal light report to graphically display key project measurements and status. Others use a simulated instrument panel (i.e., the "project control panel") to show various elements of the project.

Consistency of reports and reporting templates between projects, divisions, and global units is important. Dissimilar reporting formats increase the difficulty in understanding the data and information, reading time is increased, and combining or rolling up the data into the master report is more cumbersome. A benefit reported by benchmarkers of project organizations is that the metrics and report templates are more consistent across the organization.

> **Standard:** **The best practices project organization has consistent reporting templates for use in measuring individual project performance.**

The most popular communication mediums used are organization wide intranets and the internet. Many benchmarking groups also utilize web pages for communicating and reporting performance. Some have custom project information data systems.

In general, benchmarkers report that they continually search for software and reporting tools that are adaptable to their specific organizational situations. For reporting project schedule performance most benchmarking organizations use off-the-shelf project management software. Performance of individual projects can be rolled up into master portfolio reports. If broader coverage is required, organizations combine various software packages such as project scheduling, accounting, reporting, and communications software. A major obstacle reported is integrating project scheduling software with organizational accounting software. At this point in time, there is no generally accepted software package that satisfies all the needs of portfolio managers in large functional organizations.

7.5.3. Physical Observation Through the Audit

Even though the project organization is working with a portfolio of projects that is geographically dispersed, and the process is difficult, benchmarkers emphasize that the best way to evaluate project progress is to physically observe and verify. All other methods of performance assessment involve second hand reports and delayed communications. Observation provides immediate feedback of project status and validates information being submitted by the project team. It gives access to a broader range of input. All team members and stakeholders can be interviewed if necessary. Nonverbal signals and clues can be appraised. At the conclusion of the process, project teams and stakeholders are provided independent and objective information.

> **Standard:** **The best practices project organization audits projects by physically observing and verifying performance.**

From a portfolio governance view, the audit serves to identify projects that are not fulfilling their original promise and should be modified or terminated. Auditing improves project performance. It ensures that project scope related to organizational strategy is maintained. Cordero's research concludes that the auditing process accelerates project speed without requiring substantial changes. Scheduling audits and other reviews focuses the project teams on meeting the goals and deliverables associated with the process. Tying audits to milestone or phase gate reviews gives additional emphasis to the performance aspects of the process.

Auditing signals that management is professional and expects accountability. For example, without observation and verification, benchmarkers report that there is a tendency for reporting to become sloppy and misleading. There may be an inclination for team members to overestimate the degree of completion of individual work packages. Hence the phrase "We are 80% complete with 80% left to perform."

Using an outside auditor or team provides a safety valve for identifying and communicating project problems to stakeholders. The auditors rather than the project team can serve as the bearer of bad news or the couriers of doom. It is an invaluable benefit. Researchers Morrison and Milliken found research examples spanning 20 years that team members know the truth about the status of projects but dare not speak the truth. They found that many organizations discourage rocking the boat by challenging management decisions and are overtly intolerant of employee dissent. They determined that less than 30% of supervisors surveyed felt their companies encouraged employees to express opinions openly. Another 70% said there was a wide range of sacred cows and forbidden topics about which they were "afraid" to talk. Examples given were poor managerial decisionmaking, organizational (e.g., project) performance and inefficiencies, as well as managerial incompetence and pay inequality. The researchers also found that top managers fear and feel threatened by negative feedback and questions about current courses of action from subordinates. The managers associate the negative feedback with embarrassment and feelings of vulnerability or incompetence.

Benchmarkers also report that the strategy minimizes the problem of shooting the messenger. As a case in point, Ingram's research uncovered 14 incidents where project managers were ostracized, removed from the job, or reassigned after having taken a stand that a project was in trouble, poorly conceived, or impractical.

In the same manner, the audit team can be used as a tool to make sensitive but necessary improvements within the project team itself. Sometimes, team members are so close to the project that they are unable to recognize changes in direction that would be readily apparent to outside observers. In other cases, individuals or groups may require negative feedback and possibly corrective action related to their performance. For the project manager to attack the problem in a direct manner, could result in reduced morale and increased internal conflict between team members. If the bad news comes from an outside source, the negative feelings of the individual or project team are directed toward the outsider (who will soon be leaving) rather than inward toward the team or individual team members.

7.5.3.1. The Audit Process

The audit is a formal examination of the project. It entails a detailed appraisal of the project group's performance and progress. It is similar in nature and purpose to other review formats such as milestones, phase gates, and go/no go decision points. The major difference is that the audit entails *visual confirmation* of project performance.

Benchmarkers describe the audit as a "project health check." Its objective is to identify symptoms of illness or potential health problems where the prescription of corrective medication will result in improvement. Just as importantly is the need to verify areas of sound health where the team is doing well. One benchmarker described the auditing process as simply an attempt to answer the question, "Is the project actually doing as well as the project team says it is doing?"

To answer the question, auditors verify that the performance factors detailed in the project plan are being performed as scheduled. Also assessed is the probability of success based on the current project status. Deliverables and key success factors are emphasized as are any items that are critical to the organization. Topics verified during the audit include the project's performance in meeting schedule, cost, and specifications, as well as being on track in achieving objectives.

Usually, the audit approach is positive in nature. The audit is treated as a learning experience. Its objective is to provide the project team with useful information to help manage the project. In some benchmarking organizations, the auditors are represented as "coaches" with the intent of making the audit and review process less intimidating.

In other cases, the objective of the audit is more hard hitting. Its underlying purpose is to uncover problems early. The earlier the problem identification, the less expensive and disruptive is the response required for correction. The audit attempts to determine if and why a project is deviating from the project plan and whether it should be terminated or modified. Sometimes critical audits are scheduled because the project falls in the "bet the company" category. In these cases it is good to have a more aggressive audit group.

Benchmarkers agree that for audits to be successful, everyone must know the rules of the game, the focus of the exercise, and what is expected. They relate that where the audit process has been implemented and used, the process is well accepted. For example, in large project based companies like construction and consulting organizations, the audit team is part of the culture. The audit process has been ingrained to the point there is audit repeatability and a standard approach. As a result, generally the project team corrects the issues before the audit team arrives.

Usually, the project team is given a rating at the conclusion of the audit. In addition to the overall rating, listed are areas of superior and acceptable performance and a series of specific recommendations for improvement. In every case, the team gets feedback and is given a chance to respond in case the audit team missed or misinterpreted something. Where problems are acknowledged, the project team is encouraged to resolve the situation. When needed, a follow-up audit can be conducted to determine if the changes and improvements have been made.

7.5.3.2. Audit Timing

Often the audit is scheduled to coincide with other impending project reviews such as milestone, phase gate, and go/no go decision points. Audits may be scheduled at periodic time periods or triggered by events. For example, the project organization would evaluate existing projects in the portfolio when there is a major technology shift, the project environment and assumptions have changed, risk events have occurred that have altered the nature of the project, or scope changes have distorted the schedule, budget, and specifications.

Benchmarkers agree that an audit of the project should be performed early in its life. Funds expended are lower and ease of making improvements is greater. Progress measurement early in the project process was recommended by 25% of the more successful project teams surveyed by Gupta and Wilemon. Multiple audits can be scheduled for large projects, projects that are strategically important, and projects that are being monitored closely to evaluate the response to prior problem identification.

7.5.3.3. Composition of the Auditors

Researcher Zangwill found that audit teams vary in composition from executive groups to specialized audit groups comprised of experts in the functional fields represented by the projects. Sometimes outsiders are utilized who had little knowledge of the background and technical aspects of the project.

Benchmarkers say that ideally, the project audit or review team is composed of independent and objective reviewers from outside the project team. In reality, most benchmarkers described audit and review teams composed of stakeholders and representatives from senior management and associated critical areas such as engineering, quality, service, marketing, production, and accounting/finance.

Audit data gathered by independent auditors is generally felt to have higher credibility than project team generated conclusions. Use of outsiders makes it easier to determine the true status of the project. For example, researcher Tom Ingram found numerous examples where the image presented by project leadership was different from the reality found by auditors.

Some benchmarking participants also include a few audit team members who know little or nothing about the specifics of the project. Some incorporate reviewers from *outside the organization*. The reasoning is that an outsider is less susceptible to cultural constraints such as, "It's the way we have always done it." Use of outsiders to conduct the audit also increases the credibility of the audit. When outsiders are used, benchmarkers stress the need to provide them a tool kit of minimum standards for the project.

For internal review purposes the project team can conduct its own audit. Most use a structured approach such as a SWOT analysis that reviews strengths, weaknesses, opportunities, and threats. When combined with a review of the plan, most teams can successfully develop a realistic appraisal of their efforts.

Since auditing originated in the accounting profession, Ingram's research found that some companies were inclined to rely upon accounting personnel to perform the project audit. One benchmarking participant hired a major accounting firm to conduct project performance audits. Ingram's investigation concluded that accounting auditors often fail to see structural and people problems. They are trained to evaluate problems related to recordkeeping and to determine that money is being spent according to the budget and accounting rules and procedures. Accounting auditors often fail to spot a project that is encountering serious non financial troubles. Even when the accounting professional is aware of the problems, they may be hesitant to disclose the problem because it falls outside the scope of their traditional numbers focus. In particular, if the problem does not conform to clear rule violations, the accounting auditor may be reluctant to raise a red flag. Ingram concluded that unless the audit is focused strictly on financial aspects, most would recommend the audit be performed with a broader spectrum of functional representation.

7.5.3.4. Measure the Value of the Audit

Like all aspects of professional project management, there should be clear benefits resulting from the audit or review process. Improvements to project processes and identification of problems should be estimated in terms of savings and/or incremental revenue generated. Each successive audit should uncover areas of improvement that can be quantified. Savings can be recorded as percentage of total budget. When failing projects are identified early and terminated, savings to the company from not continuing the project can be calculated. This information serves to validate the value of the review or audit.

7.5.4. Motivational Aspects of Monitoring and Oversight

Benchmarkers relate that portfolio management can motivate all projects to improve their performance. Peer pressure can be used to maintain schedules and meet performance targets. The process involves publicly recognizing the status of all the projects in the portfolio. Schedules and individual performance can be posted and ranked.

7.5.5. Monitoring and Oversight Problems

Again it is necessary to remember that the emphasis of the project manager has shifted from development of the project charter and plan to getting the work done through people-focused leadership and management. Provision of data to the project organization is an intrusion on this process and slows the speed based elements of the project manager's efforts. Consequently, benchmarkers advise that portfolio management should request the minimum information needed to provide effective oversight of the project. Excessive detail and rigor of the methodology process can become a distraction for the project teams and slow achievement of the project goals and objectives. Benchmarkers report that sometimes the rigor of the reporting process becomes so stringent and burdensome that it overshadows the goals and objectives of the projects. Some reporting systems have even failed because of the difficulty and expense of gathering and maintaining the measurement data.

Standard: **The best practices project organization requests the minimum information needed from the project teams to ensure oversight of project performance.**

An associated problem reported is the difficulty in maintaining a high level of reporting speed. Often the need to fill out and submit status reports is low on the priority list of the project team, particularly if they are under pressure to meet performance deadlines. Usually the best solution is to have a specific person who is not involved in the higher priority elements of the project assigned to prepare and submit the reports.

References: Morrison and Milliken, 2000; Ingram, 1998; Zangwill, 1992; Cordero, 1991; Gupta and Wilemon, 1990.

7.6. STRATEGIC ASPECTS OF PROJECT TERMINATION

Terminating projects in the portfolio involves two strategic topics: the first entails the need to encourage prompt delivery of each project's products or services to the customer. The second involves the necessity of prematurely terminating projects that have gone astray or no longer fulfill a strategic purpose.

7.6.1. Encouraging Project Delivery

All efforts of the project organization and teams culminate at one point, the presentation of the deliverables, or the delivery of the product or service to the customer. No matter how well other aspects of the portfolio are performed, if projects never deliver their product or service to the customer, the project is unsuccessful. Consequently, a key element of portfolio management is making sure that the delivery is made.

> **Standard:** **The best practices project organization emphasizes expedient delivery of the project's product or service.**

Although it sounds like a simple and logical conclusion to a long process, benchmarkers report that delivering and terminating the project is a problem in most organizations. Surveys of benchmarking organizations indicate that many projects take on a life of their on and go own forever. At one benchmarking forum attended by 21 high tech companies, *every* company indicated that there were numerous projects within their companies that never terminated. Another company conducted an inventory of their project portfolio. The evaluation disclosed that 5% of projects had no foreseeable termination. In complex information systems projects, benchmarkers described a tendency to have an unclear transition defined between systems *development* and systems *maintenance*. The result was that formal handoff of the system to the customer never occurred. Benchmarkers working with new product development projects said that there was often an inclination to continually incorporate technological improvements and to improve existing designs.

The role of the project organization is to provide oversight and motivation to get the projects delivered and terminated. For information systems projects the teams can be encouraged to formally conclude the development phase of the project with a termination ceremony or celebration. It then "officially" commences the maintenance phase. For high technology or projects where the design is constantly changing, the project organization can enforce the "shooting of the engineer" concept. The scope of the project is locked in and a firm termination date is set. No subsequent scope change recommendations are entertained.

7.6.2. Terminating Troubled Projects

The second group of projects subject to termination are those that are in trouble, failing to meet the promises presented in the project charter, no longer achieve the organization's objectives, or are in situations where the environment or project assumptions have changed.

Standard: **Best practices project organizations have a formal process to terminate marginal and failing projects.**

Benchmarkers report that the dynamics of putting a project to death are incompatible with most organizational cultures. Ingram's research supports the benchmarker's experience. He found that stakeholders are reluctant to step forward and declare the project a failure or even in trouble. In cases where *everyone* privately recognizes the problems and knows the project is a terminal issue, the project in question continues ad infinitum down the path of disaster. The reality is that canceling a project is considered to be failure in many organizational cultures. The research of Ingram also found that project managers internally evaluate the *personal utility* or impact of keeping a clearly failing project ongoing versus terminating it or even making its problems public. Often the project manager concludes that there is less negative impact upon their careers if the project continues and the problems are swept under the carpet.

At the other end of the bell curve are companies in the benchmarking group that excel at the termination process. In these best practice cases, the termination process is a key element in influencing the strategic success of the host organization. For example, all pharmaceutical companies in the Benchmarking Forum had well-developed processes to quickly review and to make go/no go decisions about new projects. The reason is that pharmaceutical companies report that only one in ten thousand compounds researched ever reaches the market. Consequently, a core success factor is to rapidly identify and discontinue projects that are questionable. The pharmaceutical companies also emphasize that they formally leave the door open to reappraise projects that have been killed. It has been their experience that new uses will often be found for compounds that were previously judged of little value. They note that many of the most popular drugs on the market were killed and then resurrected by the evaluation process.

Because of the difficulty in informally terminating projects in most organizational cultures, benchmarking organizations almost always use the formal auditing process to determine whether to cease projects. Often the decision process is tied to scheduled milestones, phase gates, and go/no go reviews.

Some benchmarking organizations use project ranking to eliminate marginal projects. Ranking metrics vary from the evaluation of strategic factors to calculating financial return and potential risk. Other benchmarking participants give an overall ranking such as high priority, medium priority, or low priority. The low priority projects are scrutinized to determine if they should be dropped from the portfolio. One benchmarking organization states that they *always* terminate the bottom projects. Their logic is that resources are in short supply and that there are perpetually prospective projects that offer far greater returns to the organization.

Some benchmarking organizations utilize performance thresholds to trigger the termination audit. For example, one participant looks critically at projects that have schedule delays or exceed the budget by over 10%. They audit the reasons for the excesses, scope changes, and whether the project is meeting its deliverable targets. Another organization includes evaluation of project risk factors. They start with an assumption that, on average, they want a 90% probability for project success. During the audit the team will make a subjective estimate of the risk of failure. It the risk of failure is 35% or higher, a termination evaluation is triggered. The project team is challenged to develop plans to mitigate the risk potential down to a target level of 10%. If the acceptable target risk level cannot be attained then the project organization starts evaluating an exit plan versus other alternatives.

7.6.3. Benefits of Rapid Project Termination

Emphasis on rapid delivery of the project product or service as well as identification of marginal projects to terminate offer numerous benefits for the organization.

7.6.3.1. Free Funds for Other Investments

By generating sales sooner, funds and resources are made available for other investments. One benchmarking organization found that it was taking over four years to "officially" recognize that a project was a failure. After implementing the audit process the time to identify and terminate projects was reduced to six months. Funds and personnel made available for other uses were estimated to be in the hundreds of millions of dollars.

7.6.3.2. Execute More Projects with the Same Resources

Increased project speed means the project team will execute more projects in any given amount of time. Where previously the project team might manage a project for, say, a year or so, now each project is completed in a few months and the team then moves on to another. As an example, the Avraham Y. Goldratt Institute posted a case study on their web site describing critical chain project management at Lucent Technologies. Not only did Lucent reduce cycle time to produce new products by 50%, but the *number of projects completed* more than tripled, from 5 projects in 1998 to 16 projects in 1999, with no increase in staff.

An associated benefit of executing more projects with the same resources is that cash invested in each project is tied up for a shorter period of time. The financial effect is similar to that obtained from increasing inventory turnover. High inventory turnover means that the investment in inventory is small compared to the amount of resultant sales. In the case of project management, resource usage is maximized by (a) making project managers and team members available for other high potential projects, and (b) freeing the financial resources that would have been allocated had the project run for the original extended period of time.

7.6.3.3. Generate Incremental Sales

The value of reducing time to market by one day is phenomenal. Each day that time to market is reduced adds one more day of sales. One benchmarking telecommunications company shortened time to market from 52 to 18 months with the result that sales occurred nearly *three years earlier* than the prior approach. The project group estimated incremental sales resulting from the 34-month reduction amounted to $4 billion.

References: Cabanis-Brewin, 2000; Ibbs and Kwak, 1997.

7.6.4. Communicating the Benefits of the Project to Stakeholders and the Host Organization

Chapter 2 emphasized the importance of building a business case that communicates the value of professional project management for the organization, senior management, other stakeholders, and individual project managers. Chapter 6 added detail by showing the manner in which performance metrics are developed and measured. Included was a listing of empirical and subjective metrics utilized by benchmark organizations during the execution and control phases of the project. Now, at project termination, the task of communicating the performance and benefits as recommended in Chapter 2 takes place.

At the termination phase of the project, communication of the individual project's and the project group's performance is crucial. When originally approving each project's charter, the host organization put into action a strategic decision. At the conclusion of the project, it is important for senior management to know whether the project fulfilled its initial promises. In addition, benchmarking participants unanimously agree that for project management to continue to be recognized as a profession of value to organizations, there must be a clear linkage between its application and improved organizational project goal achievement.

Despite the generally recognized importance of such information, historically, project management groups have been slow to quantify and measure the value of professional project management in their organizations. Ibbs's research determined that few project groups measure the benefits to the organization of individual projects and the project organization. One reason is that best practices project groups have been measuring project performance throughout the execution phases and sometimes assume that stakeholders are familiar with the overall performance of the project. Benchmarkers relate that, because the benefits seem so clear to team members and the project organization, often the assumption is made that everyone else also recognizes the value of project management. This is seldom true. Measurement and communication of benefits is further restricted because those who do set out to measure and communicate final project benefits find that measuring the results of project management is easy to talk about but hard to do.

As a result of these experiences, there is conviction among forum participants that the bottom line as well as subjective and generalized benefits of the project management must be identified, quantified where possible, and communicated to stakeholders, senior management, and other functional areas of the organization.

Standard: **The best practices project organization measures and communicates project contributions to stakeholders and the enterprise.**

As always, there are political and cultural problems related to measuring and communicating project success. It is a fine line between the appearance of egotistic bragging and the need to publicize positive results and contributions to the organization. Most would suggest emphasizing the contributions of stakeholders, vendors, team members, and associated functional areas. Benchmarkers also stress that sometimes the results attained that are the most valuable to the organization are the subjects that can't be talked about, i.e., saving the losing projects of other groups.

In addition to all the measurement metrics detailed in Chapter 6, additional approaches organizations take to communicate project benefits follow. Any or all of these measurement methods can be tailored to articulate project contributions with the host organization's goals.

7.6.4.1. Compare with Prior Projects or Approaches

The performance of the project team can be compared with historical projects. If sufficient historical examples are available, cost and performance trend lines can be developed and compared with the results of the current project. Where specific metrics are not available, the current project can be subjectively compared with the way projects were managed in the past.

7.6.4.2. Client Retention

The proof of customer satisfaction is retention as evidenced by repeat business and strengthening of the client relationship.

7.6.4.3. Stakeholder Satisfaction

When taking a stakeholder view of team performance, success becomes a multidimensional issue. Issues in addition to the project's success can be taken into account. Different constituencies have varying concepts and definitions of superior performance. The opinions of interest groups and individuals that could impact the perception of project success can be evaluated. Where there is evidence of satisfaction, this should be communicated.

7.6.4.4. Reduction in Time to Market or Completion Time

When a new product is involved, the practical benefit of reducing project time is that the product enters the market sooner. Generally, one of the most dramatic results of professional project management is the reduction in time to complete projects. As a result, one measurement of project performance is to compare the former time to complete projects, or "lead time to market," with those achieved by the current project. Generally, the historical time to develop new products is well known within an organization. It is rarely precisely quantified but experienced managers and executives usually have a general idea of traditional lead time to market.

7.6.4.5. Measure the Impact of Problem Identification and Corrective Action

One advantage of continuous measurement and monitoring of project performance is that marginal projects are identified and culled earlier in the development process. Time to termination can be logged and the results compared with historical efforts. Other groups report that they measure the time to identify problems and the number of project problems corrected.

References: Ibbs and Kwak, 1997; Ingram, 1998; Ancona and Caldwell, 1990.

7.7. CONCLUSION AND MEASUREMENT
OF PROJECT ORGANIZATION SUPPORT
OF ORGANIZATIONAL STRATEGY

An evaluation of a project organization's level of support for the host organization's strategy can be numerically quantified using the orga-

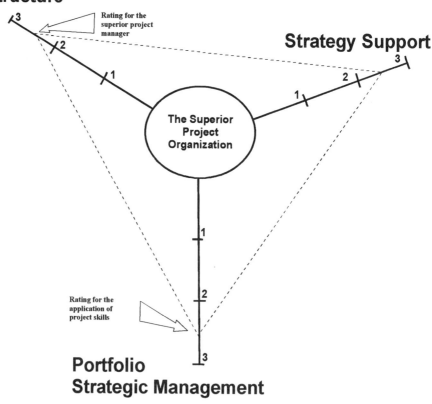

FIGURE 7.1

nizational support competencies subwheel. The numerical results from this wheel are rolled up or transferred to the master project organization competency wheel. In Figure 7.1, it can be seen that the project organization's support of the host organization's strategy is comprised of three elements: the implementation of an efficient project organizational structure; implementation of the host organization's vision, strategy, and goals; and the strategic management of the portfolio of projects. Each of these are allocated equal weighting in terms of their impact on project goal achievement.

Ranking Scale

Frequently, if not always	Fairly often	Sometimes	Once in awhile	Not at all
5	4	3	2	1

EVALUATION QUESTIONS

Efficient Project Organization Structure

5 4 3 2 1   1. Project goal achievement probability is maximized because the project manager, project office organization, host organization, and external environment components work in unity to achieve common objectives.

5 4 3 2 1   2. Stakeholders recognize the superior project manager as the single most important factor impacting project goal achievement.

5 4 3 2 1   3. The project organization builds partnerships with and gains support from senior executives.

5 4 3 2 1   4. The project organization has a formal presentation that outlines the benefits of the project organization.

5 4 3 2 1   5. The project organization proactively communicates the benefits of the project organization to senior management and other stakeholders.

5 4 3 2 1   6. The project group structure is compatible with and supports the organizational strategy and structure.

5 4 3 2 1   7. The project organization reports to a multifunctional executive or executive committee.

Supporting Organizational Strategy

5 4 3 2 1   8. The project organization directs all activities to maximize the probability of achieving the host organization's goals.

5 4 3 2 1   9. The project organization aligns and integrates vision, goals, and plans of individual projects with those of the host organization.

5 4 3 2 1   10. The project organization sets clear, measurable, and observable goals.

5 4 3 2 1   11. The project organization maintains constant focus on the host organization goals as supported by the project organization and specific project goals.

5 4 3 2 1   12. Professional project management is a core strategic process within the host organization.

5 4 3 2 1   13. The project organization is actively involved in the project management process from inception to termination.

Managing the Strategic Project Portfolio

5 4 3 2 1   14. The project organization manages projects as a coordinated portfolio.

5 4 3 2 1   15. The project organization clearly defines and communicate that ultimate responsibility and accountability for the project rests with the project client or owner.

5 4 3 2 1   16. The project organization maintains a database of all major projects in the host organization.

5 4 3 2 1   17. The project organization selects projects by evaluating and ranking strategic factors, financial benefits, risk, and other appropriate factors.

5 4 3 2 1   18. The project organization monitors and supports the project plan process to ensure that the plan satisfies the strategic objectives set forth in the charter.

5 4 3 2 1   19. The project organization has clear, easy to apply and understand metrics against which project performance and progress are measured.

5 4 3 2 1   20. The project organization measures project performance by evaluating customer satisfaction.

5 4 3 2 1   21. The project organization defines critical success factors in measuring project goal achievement.

5 4 3 2 1   22. The project organization aggregates project goal achievement and/or benefits of projects.

5 4 3 2 1   23. The project organization is on time and on budget.

5 4 3 2 1   24. The project organization requests the minimum amount of reporting needed from the projects to effectively monitor portfolio performance.

5 4 3 2 1   25. The project organization creates a virtual project office environment that duplicates the benefits of collocation.

5 4 3 2 1   26. The project organization selects superior project managers.

5 4 3 2 1   27. The project organization manages project scope changes with a formal scope change process.

5 4 3 2 1   28. The project organization monitors the project portfolio in sufficient detail that no single project deviates materially from its original project plan without portfolio management being aware of the deviation.

5 4 3 2 1   29. The project organization has a formal monitoring and oversight process for the project portfolio.

5 4 3 2 1   30. The project organization uses variance analysis as a tool to monitor and control deviations from the project plan.

5 4 3 2 1   31. The project organization has consistent reporting templates for use in measuring individual project performance.

5 4 3 2 1   32. The project organization audits projects by physically observing and verifying performance.

5 4 3 2 1   33. The project organization requests the minimum information needed from the project teams to ensure oversight of project performance.

5 4 3 2 1   34. The project organization emphasizes expedient delivery of the project's product or service.

5 4 3 2 1   35. The project organization has a formal process to terminate marginal and failing projects.

5 4 3 2 1   36. The project organization measures and communicates project contributions to stakeholders and the enterprise.

To determine the number of points to apply to the master organizational competency wheel, score the questions as follows:

	Total self-rating	Maximum
Questions 1–7: multiply the points selected by 0.10	____	3.5
Questions 8–13: add the points and multiply by 0.117	____	3.51
Questions 14–36: add the points and multiply by 0.03	____	3.45

Part III

Providing Support for Project Goal Achievement

Part II described the project organization's role as administrator and supporter of the *host enterprise's* vision, goals, and strategy. Part III addresses the numerous ways the project organization provides support for the *individual project teams* in achieving project goals. As depicted in Figure III.1, the project organization is structured for and stresses speed, efficiency, and effectiveness. Included is emphasis on multifunctional teams. Knowledge is managed through the provision of standardized and predictable methodologies and approaches to managing projects. Also provided are enterprise wide tools, templates, and expertise. A knowledge library transfers information from one project to the next and makes available models upon which to pattern new project charters and plans. Nurtured is competence building training that focuses on project specific skills, professionalism, and character. Project support maximizes people performance. Some benchmarkers consider project team member selection and people motivation and guidance the most important performance elements of project management. Compensation methods are important motivators as well as other forms of nonfinancial rewards. Provided is ambassadorship capability for promoting project interests, smoothing cultural and political issues and maintaining the host organization's confidence.

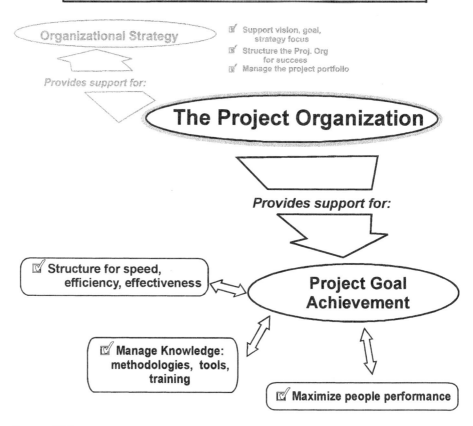

FIGURE III.1

8

Managing for Goal-Achieving Speed, Efficiency, and Effectiveness

The superior project organization promotes speed based project management. It encourages projects be performed as speedily as possible while maintaining efficiency and effectiveness. The overall objective of speed based project management is to have the project team complete the project in as short a time as possible and then to have them move on to another project. The benefits are enormous. Lead time to project completion is reduced. Products and services generate sales and positive results sooner, market dominance is enhanced, resources are freed for other investments, and more projects are completed with the same amount of money.

An analogy of the application of speed, efficiency, and effectiveness is to envision the task of an individual desiring to travel from Los Angeles to New York City. The *speed* aspect of the task means that the individual will fly in a jet rather than drive an automobile or ride a train. To be *efficient* the person would take the most direct, non stop route instead of a circuitous route with numerous stops and aircraft changes. The traveler becomes *effective* when the goal of reaching the destination has been attained. In like manner, the superior project manager is constantly cognizant of the necessity for speed while maintaining efficiency and effectiveness.

Standard: **The best practices project organization emphasizes project speed in achieving goals, while maintaining efficiency and effectiveness.**

8.1. SPEED

The overshadowing aspect of speed permeates nearly all elements of superior project management. Surveys of project customers support that the most important performance measurement of project success is meeting time targets. Time is consistently rated as more critical than meeting cost targets and satisfying technical and functional specification requirements.

Speed consciousness serves as a natural restraint against the tendency for organizations and people to add increasing amounts of complexity, bureaucracy, formality, methodology, use of templates, project software, and planning. The speed factor forces recognition that excess in any of these areas will slow the project and that there is a constant need to efficiently and effectively achieve the project goal.

Maidique and Zirger established that companies with short product development times consistently outperform those with longer cycles. The first product into a high demand market can command a premium price as well as establish a market niche and respond to rapidly changing customer demands according to the supporting research of Hopkins. A McKinsey and Company model was more specific. It showed that high tech products entering the market six months late earn 33% less profit *even though they came in on budget.* The projects in the study that came in on time and 50% over budget reduced profits only 4%.

Project speed offers many benefits for the host organization. Being first to market amplifies competitiveness. It results in faster response to market needs, higher market share, and access to higher profit margins. Bringing the project in on schedule results in happier customers and offers greater potential for additional future business. Managing for speed reduces product cost. By increasing the number of projects completed by the project team, return related to project team cost is multiplied. Rapid technological change and the resultant product obsolescence and shortened product life cycles also encourage faster project completion. Corporate growth objectives, global strategic initiatives, and pressure from senior management are added to result in the conclusion that the superior project team has no choice but to reduce project end-to-end cycle time.

Superior project teams are found in organizations where speed is a central objective of the firm and part of its mission and strategy, according to Cordero's research. His data suggest that increased project speed combines positive qualitative estimates with traditional cash flow and capital justification. Gupta and Wilemon show the value of speed when it is an integral part of the organizational culture. They also show that speed is a difficult concept for many organizations to embrace. In their study, 53% of respondents feel their organizations are unwilling to take the appropriate steps necessary to shorten development time. Excessive red tape, organizational inertia, risk aversiveness, conservatism, and low priorities for project based programs are obstacles reported.

Just as in driving a car, there are dangers associated with speed. The project manager is required to maintain a high state of alertness while applying superior piloting skills. When there are problems, they occur at a faster rate, can cause more damage to the project and associated stakeholders, and demand all the superior project manager's capabilities to control and mitigate. For these reasons, it is stressed that the speed aspect of superior project management *must be* accompanied by efficiency and effectiveness.

References: Cordero, 1991; Eisenhardt and Tabrizi, 1990; Gupta and Wilemon, 1990; Maidique and Zirger, 1984; Hopkins, 1980.

8.2. EFFICIENCY

The superior project organization encourages project teams to take the shortest and most direct route between two points. The specified project activities, tasks, and deliverables, *and no more*, are completed and delivered. Fulfilled are the functional, technical, and quality specifications for the project without adding unneeded extras. Utilized are the most simple and least complex amount of methodology, formality, procedures, and templates necessary to speedily and effectively attain the project goal. In general, researcher Galbraith concluded that when situations are certain and predictable, there is greater emphasis on planning and organizing. There is also greater

reliance on routine and bureaucratic and mechanistic teams and organizations. In conditions of high uncertainty, a more flexible, adaptable structure and style is appropriate.

Efficiency problems in project organizations are observable. It is
common for benchmarking participants to relate that their organizations initially adopted the attitude that if a little methodology, a few
templates, and some project structure are good, then *a lot* should be
even better. Some have generated and attempted to implement multiple volumes of project management procedures and methodologies.
One university with a consulting arm has widely promoted and has
gained acceptance of a project organization maturity model based on
increasing amounts of formality, procedures, and bureaucracy, and
in which the project manager is considered a less important consideration. The result of all of these approaches, invariably, is a tools and
procedures dominated project approach that is slower and more cumbersome than a less burdened approach.

Guidance and support by the project organization is beneficial
because project managers can also suffer inefficiency handicaps. Reports are common about project managers who spend such a large
proportion of their time on the computer updating and manipulating
the project schedule, or filling out reports and other associated documentation, that they fail to manage the people on the project. For
example, one of the questions forum participants answer is, "What
percentage of your time do you spend on the *tools* of project management?" (i.e., the software, methodology, templates, etc.). Answers
ranged from 25 to 75% with even distribution over the entire range.
A basic statistics text would indicate that such a result is an indicator
of a survey anomaly. Consequently, a control question was added
which asked, "What percentage of time do you spend on *face-to-face
management* of projects?" Again the responses ranged from 25 to 75%
of the time with even distribution over the entire range. Clearly there
was an unknown factor involved since the two questions were diamet-
rically opposed to each other. After discussion at several benchmarking forums, it was concluded that project managers early in their careers tend to spend more time on the tools of project management;
in fact as much as 75% of their time according to the survey results.
As they become more comfortable with the tools and the project management process they exert more time on the face-to-face aspects;
again, as much as 75% of their time. There was general agreement
that the face-to-face management style used by more experienced
project managers resulted in greater efficiency. The conclusion of
benchmark forum participants was that as the typical project manager becomes more proficient and advances along the learning curve,

he or she naturally become more efficient at the project managing process. The project organization can encourage this development through training and mentoring.

Project organizations should encourage an appropriate amount of planning. Eisenhart and Tabrizi's studies of fast-moving, volatile high technology projects showed that excessive use of planning and software slowed down the project and reduced effectiveness. In the tests, *planning resulted in slower product development time* as did increased supplier involvement. Use of software also slowed product development. Utilization of cross-functional teams consistently resulted in faster product development as did overlapping development phases, powerful team leaders, and frequent testing combined with higher numbers of design iterations. Face-to-face interaction, flexibility, rapid learning, and improvisation were considered essential by the most effective and fastest teams.

Eisenach speculates that the overemphasis on software slows projects in uncertain situations because is takes a long time to learn how to use it effectively. Also, some firms tested used incompatible software programs. Further, some team members became so enamored with and addicted to the software that they lost focus on the project itself. Others stated that sometimes software simply automates well-known procedures and old approaches. While they are applicable to predictable projects, they should receive less emphasis when working on projects that are changing often.

Overemphasis on planning for uncertain projects slows down developers and teams who could move more quickly. One explanation given is that for high velocity projects that change rapidly and unpredictably, extensive planning is simply a poor use of time. Planning may also slow down the project pace when information is obsolete or incomplete.

Note that it is emphasized by benchmarking participants that people new to *the project management profession* should continue to rely upon planning and the methodology and software tools of project management. *The schools and training organizations* should continue to emphasize project management tools. As usage of tools, methodologies, templates, and procedures becomes familiar, then the superior project manager will naturally devote more time to people management and accelerating project speed. It is also emphasized that the more predictable and repeatable the project, the more it lends itself to the use of extensive planning and use of project software tools.

References: Eisenhardt and Tabrizi, 1990; Galbraith, 1973.

8.3. EFFECTIVENESS

No matter how fast and efficient the project team, if it doesn't reach
its goal, it is considered by most to be a failure. Using the flight from
Los Angeles to New York as an example, the aircraft could be flying
at record speeds with a good tail wind and taking the most direct and
efficient route, but if it lands at the wrong airport either through design or because of bad weather, error, or emergency, most of the customers will likely be unhappy with the outcome. In project management, the superior project manager attains the goal, terminates the
project, and moves on to the next.

As discussed in prior chapters, a primary function of the project
organization is ensuring that projects are delivered on time and the
project team promptly moves on to another project.

8.4. SPEED, EFFICIENCY, AND EFFECTIVENESS RELATED BEST PRACTICES

Many observable and measurable best practices in project management relate to speed, efficiency, and effectiveness. Prior chapters discussed the importance of properly structuring the project organization. Project teams as well can be structured for speed, efficiency, and
effectiveness as can the manner in which projects are planned, executed and terminated. For example, the multi functional project approach and fast tracking represent an effort to accelerate the project.
Leadership style is also impacted and tends to lend itself to "expert"
leadership, rapid decisionmaking, and delegation of authority and
control to team members. Emphasis on planning and monitoring of
project schedules during project implementation is a key success vari-
able. Milestone goals, periodic reviews, and audits speed the project
process by emphasizing near-term goals as well as resulting in arm's
length and objective suggestions for improvement or even the elimination of time- and resource-consuming marginal projects.

References: Millison et al., 1995; Cordero, 1991.

8.4.1. Project Organization Structure

Chapter 3 described the various types of project organizations utilized by enterprises in the benchmarking forum. Benchmarkers
stress that the particular type of organization selected should increase project speed, efficiency, and effectiveness within the host organization. It should be designed to give improved administration

and support of the host organization's vision, goals, and strategic plan, while at the same time provide support for the project teams. For example, the overall objective of the Project Management Center of Excellence is to advance the capability of the host organization to execute projects. The Project Support Office improves project speed by providing tools and services needed by the project teams. The Program Management Office manages the portfolio of managements with emphasis on speed, efficiency and effectiveness. The Project Office governs individual projects and applies all the speed based project management concepts. (See Figure 8.1.)

Benchmarkers caution that improvement in project speed is not an automatic outcome of the project organization. Sometimes the project organization slows performance rather than enhances it. In

FIGURE 8.1

certain cases, the project organization represents an intrusion upon project team performance. Reporting requirements are excessive rather than requesting the minimum information needed to provide oversight and give an adequate picture of project performance. Methodologies, tools, templates, and communications capabilities become so detailed and rigid that flexibility is lost and time is consumed meeting an unrealistic checklist approach to project management. Benchmarkers say that all activities performed by the project organization and all tools provided should clearly improve performance in a cost effective manner.

8.4.2. Project Team Structure

The earliest organizational researchers recognized the importance of organizing or, specifically, creating the structure and lines of authority and responsibility for the team. The concept is observable in superior sporting teams. They are structured to capitalize on the skills of the players and organized around a series of preplanned "plays."

> **Standard: The best practices project organization structures the project teams for speed, efficiency, and effectiveness.**

Usually team structure is a reflection of the project leader and tempered by the nature of the project and the environment in which it is set. The following subsections list generic team structures and a discussion of each.

8.4.1.1. The Multifunctional Team Structure

A fundamental way to improve project speed and efficiency is through the multi- or cross-functional project team structure. Many of the phenomenal increases in revenue and reductions in cost associated with projects are the result of utilizing multifunctional project teams.

By definition, multifunctional project teams consist of representatives from each functional area that are impacted by the project. For example, the new product development project shown in Figure 8.2, includes design engineers, production or manufacturing, marketing and finance personnel. The multifunctional team members work simultaneously, in a parallel effort, on each of the functional aspects of the project. Discarded is the traditional sequential approach where each functional area performs its activity and then passes the project

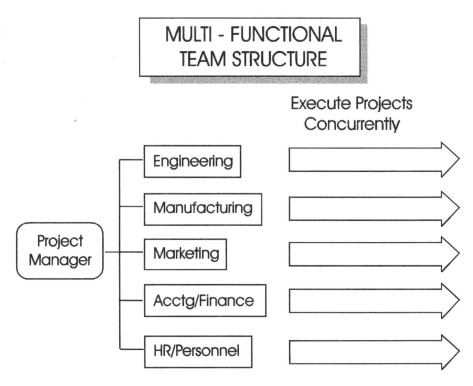

FIGURE 8.2

to the next. The primary output of multifunctional teams is *increased speed* compared to performing projects in a sequential manner.

In essence, the multifunctional team is a miniature company with the project manager serving the role of chief executive officer. It is one reason why several benchmarking organizations are using project management as a training ground for future executives. Their view is that the skills needed to lead and manage a multifunctional team are the same as required to govern the entire organization.

Clark's research team concluded that the multifunctional project team is *the* critical element in achieving project objectives. Maidique and Zirger agreed and said that multifunctional teams result in smoother execution of all phases of the development process and is the major factor in the rapid attainment of project goals. In Gupta and Wilemon's study, 33% of respondents reported that the multi functional project approach is the primary means to improve project speed.

Phenomenal arisen have resulted from the multifunctional approach. One company in the benchmarking group reduced lead time to market from 52 months to 18 months and generated an estimated $4 billion in incremental sales. Cross-functional teams have cut development time by 20 to 70%, boosted white-collar productivity 20 to 100%, and increased sales 5 to 50 times, details the research of Zangwill. Eisenhardt and Tabrizi report an almost perfect positive correlation of .87 between the use of multifunctional teams and timely project goal achievement.

8.4.2.1.1. Include Suppliers and Vendors.

Gupta and Wilemon found that the multifunctional approach was particularly beneficial when applied to suppliers and vendors. The researchers' investigation of product development projects in heavy industry determined that the fastest moving projects actively involved suppliers in the process. Malbert's research team also concluded that strong formal ties to suppliers improved the project process. Effective project leaders delegated project steps to suppliers. Lorenz identified specific areas where supplier involvement improved project goal achievement. Supplier involvement during development reduced the workload on the primary team. Supplier involvement let the primary team focus on the tasks and skills that maximized their core competencies and skills. Suppliers applied their own skills to result in the maximization of synergy. Suppliers also served as outside sources of critique and elimination of future problems as well as suggestions to improve process and product. Suppliers were highly skilled in specialized areas which gave them the capability to fulfill sudden and unusual requests promptly.

Early involvement of suppliers improved the outcomes of the design process judged the research team of Clark, Chew, and Fujimoto. There was a positive correlation between supplier involvement and project success, found Brown and Eisenhardt. They judged that suppliers and customers are key players in the project process. Supplier involvement can alert the team to potential downstream problems early on when they are easier to fix.

One benchmarking organization made suppliers a key component in the project plan for a $3 billion global communications project. It was important that the project come in on budget. Suppliers were involved from the beginning and were required to make a commitment to hold costs to their original estimate and to avoid scope changes. If one supplier came in over cost, the others (including the managing partner) were committing to absorb the cost of the change. The process required a high degree of coordination and communication but was considered successful in maintaining cost while attaining its technical and functional results.

References: Brown and Eisenhardt, 1995; Mabert et al., 1992; Lorenz, 1990, Clark et al., 1987.

8.4.2.1.2. Multifunctional Team Benefits. In addition to the benefits resulting from parallel rather than sequential efforts, other gains accrue from the multifunctional approach.

- **Problem recognition.** Several researchers have observed that multifunctional teams recognize current and down stream problems and mistakes earlier than other forms of team structure. Each functional team member ensures that problems that could occur in the team member's department are avoided early in the process. As a consequence, the problems tend to be smaller, easier, and less expensive to correct. Later scope changes are minimized.
- **Improved communications.** Representation from more organizational functional areas results in better communications. The organizational and knowledge diversity of the multifunctional team increases the amount and variety of information available to the team, both between members of the team as well as with other functional areas.

  Team members tend to communicate more with outsiders who have similar functional backgrounds. Consequently, the more functions on the team, the broader and more comprehensive the external communications. Dougherty's research showed that on multifunctional projects, the team members combine their views in a more highly interactive fashion with the result that information content is increased.
- **Increased asset turnover.** Conducting projects in parallel rather than sequentially means that the team completes more projects in a given amount of time. Project team cost related to output is minimized.
- **More compatible with overlapping design phases.** Involving multifunctional representatives early in the process reduces wait and slack time between phases and activities. Cross-functional teams permit overlap of project phases since each functional team member or group is working on its subproject at its own pace. Consequently, it is common for each project functional group to progress at different rates.

8.4.2.1.3. Dangers of Multifunctional Teams. Like many aspects of superior project management, there are inherent dangers associated with the multifunctional approach: the dramatic increases in project speed necessitate heightened project leader vigilance and skill; project time performance is harder to measure since each functional team element is likely working at a different pace; often key functional elements of the project team are not included as a result of cultural influences, and expertise of team members may decline or be limited compared to traditional functional approaches.

- **Increased need for coordination.** Multifunctional teams are fast. They often are working at three to four times the speed of a sequentially phased project. The differing rates of progress by the functional team members complicate the role of the project manager and require careful coordination and communication.
- **Harder to measure performance.** The sequential project approach of working toward total project phase gates, go/no go decision points or project milestones make it easy to measure time performance. Either the project phase is completed on the scheduled date or it is early or late. The variance in such cases is measured in units of time (for example, days, hours, months early or late).

    Overlapping activities and project phases make overall project time performance more difficult to measure and present clearly to stakeholders. Since each of the team functions could be working at a different pace, it is likely that the functional elements will arrive at overall project milestones at different times. To effectively measure time performance for the project necessitates use of more conceptually difficult non–time based performance measurements such as percentage of work packages complete, percentage of funds expended or an earned value variance measure (i.e., budgeted cost of work performed compared to budgeted cost of work scheduled).
- **Some functional areas are overlooked.** Companies in the benchmarking forum report that there is often a cultural bias toward excluding specific functional areas from the project team. Sometimes companies are not even aware that key functional elements are being excluded. Nonaka's research indicated benefits from involving often overlooked vendors at the early stages of the project. He found that understanding of goals and project objectives was improved as was loyalty and trust. In another situation, a major manufacturing organization reported that their project teams were truly multifunctional. The teams included members from engineering, manufacturing and service support. When asked why finance and marketing were not involved the response was, "Why would we want to include them?" After discussion the company representatives realized that finance is a key success

component because, like a CEO, to achieve success the project manager needs to control the money. And, of course, marketing is key because the initial sale generates all other activities. The two key functional areas had been excluded simply because "that's the way we have always done it."

- **Individual expertise may decline.** Multifunctional project teams typically recruit and obtain team members from their respective functional organizational areas. The physical detachment of the individual from the technical area of expertise sometimes makes it more difficult for professionals to maintain currency with developments in their field as well as within the company. Marquis and Straight found examples on highly technical projects where multifunctional teams recorded poorer performance than would occur in the highly specialized functional areas.

References: Jassawalla and Sashittal, 1999; Donnellon 1996; Liedtka, 1996; Kahn, 1996; Brown and Eisenhardt, 1995; Ancona and Caldwell, 1992b; Zangwill, 1992; Dougherty, 1990; Cordero, 1991; Dougherty, 1990; Eisenhardt and Tabrizi, 1990; Lorenz, 1990; Nonaka, 1990; Stalk and Hout, 1990; Clark et al., 1987; Maidique and Zirger, 1990; Clark and Fujimoto, 1987; Gold, 1987; Marquis and Straight, 1965.

8.4.1.2. The Matrix Team Structure

The distinguishing characteristics of the matrix project organization are two bosses and borrowed resources. The structure is based on appointing a project manager from within a functional area. Usually the project manager has a full time job and is reporting to a functional manager. Team members and financial resources are "borrowed" from other departments. Team members have two bosses, the project manager and their functional unit managers. (See Figure 8.3.)

The industry defines two types of matrix organizations, the weak and the strong. The difference relates to the amount of authority held by the project manager. According to the research, the strong form of matrix management would consistently have the highest probability for goal achieving success.

For example, assume that the functional manager informs one of the individuals within the department that they have been appointed project manager for the holiday party. Included on the team are a few employees borrowed from other departments. Under the *weak form* of matrix management, the functional manager might direct the project manager to find two or three restaurants. However, the func-

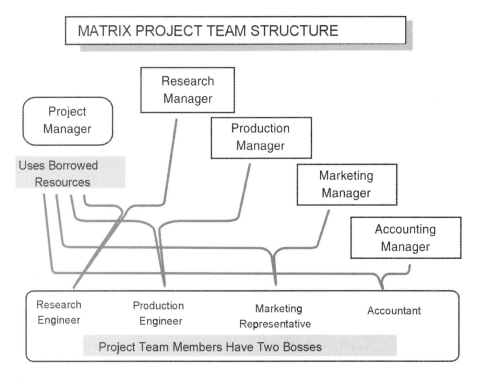

FIGURE 8.3

tional manager would instruct the project manager not to make a selection. That capacity would be reserved for the functional manager. The functional manager might then ask the project manager to obtain the names of two or three bands. Again the functional manager would reserve the right to select the band to be hired.

Under the *strong form* of matrix organization, the functional manager would assign the project manager the task of managing the holiday party. Ensuing communications would include the amount of the budget, instructions to "take care of all the details," and a request to be informed of the date and time to show up for the party.

In either case, the governance of the matrix structured project represents challenges to its manager. The most effective project manager will use finesse and persuasion in motivating people to devote effort to the project. In essence, team members follow the project manager because of leadership skills rather than the strength found in formal authority.

The matrix organization is primarily found within functional departments on projects too small to justify a full time project manager. In fact, for small projects it is the most popular and prevailing type of project organization used in large functional organizations. Often project managers will manage several matrix style projects simultaneously. At the other end of the project size scale, experience has shown that the effectiveness of matrix management on large, multifunctional projects is limited compared to using a full time project manager reporting to a higher level of the organization.

The matrix organization was popular and attained fad status in the 1970s and 1980s. Its theoretical benefits appeared so inviting that it was applied by organizations to the management of nearly all large multifunctional projects within the organization. Companies embraced the concept in hope of getting something for nothing. They turned to the matrix project organization as a way to use existing people and resources to execute projects within the cultural confines and constraints of traditionally functional organized enterprises.

The desired results rarely materialized. Organizations found that executing large multifunctional projects with a matrix structured project team, within functional departments, often resulted in an imposition on employees who were already performing a full-time job. Particularly in performance based organizations such as manufacturing or sales, applying the matrix managed projects within the organization reduced the measurable output of the group. The "borrowed resources" element of the matrix equation also proved difficult, particularly when the resources that the project team was trying to borrow were scarce funds or people.

David Cleland in his book, *Project Management Strategic Design and Implementation,* reviews the research and literature associated with matrix project organizations. He quotes Texas Instruments as announcing that their attempt at matrix management was a key reason for their economic decline. Xerox said that matrix management proved to be a deterrent to product development. Other cases cited concluded that the matrix style of project organization was more cumbersome and costly than an independent project organization.

Research quoted by Cleland supports that the more authority that resides with the project manager, the higher the probability of project success. Benchmarkers agree and relate experiences concluding that using matrix organizations to execute large multifunctional projects is consistently less successful than utilizing full time project managers and team members.

References: Cleland, 1999.

8.4.2.3. The Isomorphic Team Structure

The isomorphic team structure refers to team structures that look like the project activity being executed. For example, a project to design a new truck might include a transmission subgroup, final drive subgroup, engine subgroup, operator's compartment subgroup, and frame subgroup. Each group would be working generally independently of each other. (see Figure 8.4.) More than likely, each group would have a leader who is an expert in the technology of the group. The management role of the project manager would be primarily to monitor schedules and milestones. Groups falling behind would be notified and prodded forward.

8.4.2.4. The Specialty Team Structure

The composition of a project where specialists and other outside experts provide services for the various project modules is titled the specialty team structure. Figure 8.5 shows a basic isomorphic structure with specialists added. The manufacturing engineer works with

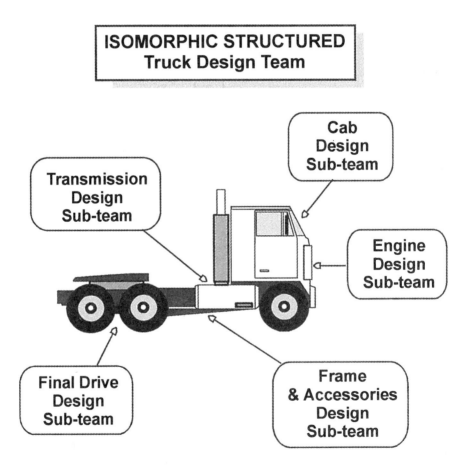

ISOMORPHIC STRUCTURED Truck Design Team

Cab Design Sub-team

Transmission Design Sub-team

Engine Design Sub-team

Final Drive Design Sub-team

Frame & Accessories Design Sub-team

FIGURE 8.4

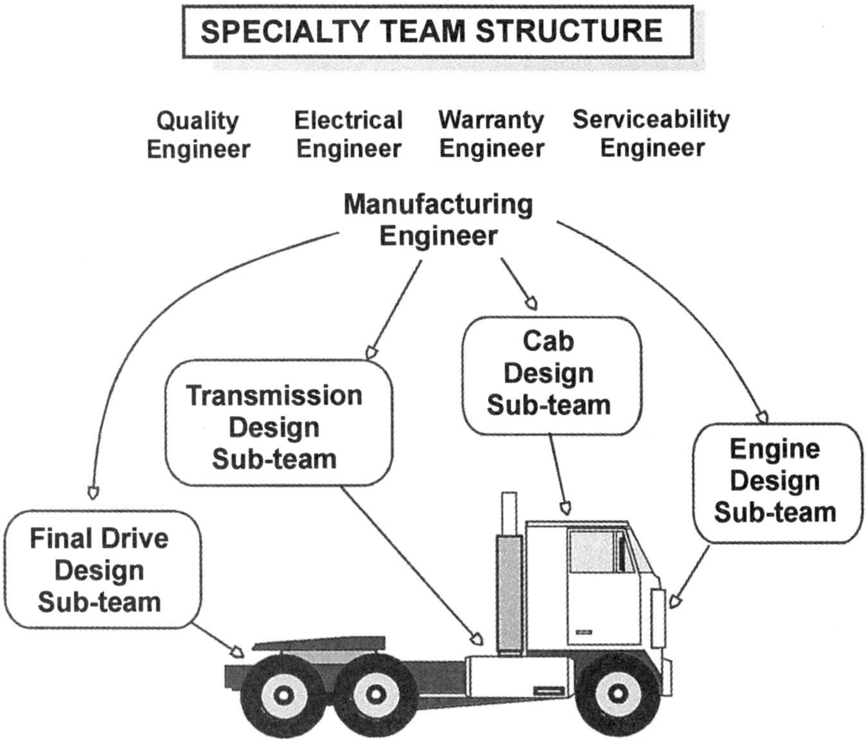

SPECIALTY TEAM STRUCTURE

Quality Engineer   Electrical Engineer   Warranty Engineer   Serviceability Engineer

Manufacturing Engineer

Cab Design Sub-team

Transmission Design Sub-team

Engine Design Sub-team

Final Drive Design Sub-team

FIGURE 8.5

each of the subteams to ensure that everything being designed is compatible with the manufacturing process. Also available to the design sub teams are other experts such as the quality, electrical, warranty, and serviceability engineers.

8.4.2.5. The Egoless Team Structure

When a team works on a project where the vision of the mission overshadows personal interests and where each individual is self-motivated, the result is the egoless team structure. (Note that the egos are still there. It's just that they are somewhat suppressed in deference to attaining the ultimate mission of the team.) Examples are service, religious, and social missions where volunteers are motivated by a higher purpose or altruism and provide services according to the skills they have to offer. For example, one of the benchmarkers related how a retired person in the local church had the shingles blown off the roof of her house during a storm. The people in the church took up a collection, bought some shingles, and reshingled the person's home. Some people did the shingling and others brought food, ran errands, or carried shingles to the roof. Everyone helped out where they could. The benchmarker said that luckily one of the people on the team was a contractor who provided expertise and managed to keep the group on track.

In organizations, egoless teams are found among experts such as engineers, professors, and certified public accountants. Such individuals have a high sense of loyalty to their professions and enthusiasm for the tasks they perform. For example, engineers might join together to solve a particularly vexing design problem. Although there would be a leader, the team members would be largely motivated by the sense of professionalism and joy of working together on a challenging problem.

Management of the egoless team typically paints a portrait of the vision and goals of the venture. They provide support for the team members in their efforts to attain the goal. Often where service organizations are involved, the individual is encouraged to develop a job around the particular skill set they can contribute. For example, homeless shelter management might include volunteers from the accounting or finance profession to do the bookkeeping, people who enjoy cooking to prepare the food, and the socially conscious could help people with individual problems. When skills are missing from the team, that particular set of activities will probably not get performed or will be performed in a minimal fashion. During all this time, the leader of the group would gently keep everyone on track and working toward the goals and fulfillment of the stated mission.

Egoless teams are, by definition, self-motivated. Even so, motivational techniques can be applied to support the process, build team spirit, and maintain high commitment levels. Inspiration for egoless teams comes primarily from self-gratification and peer recognition. Often egoless teams are composed of unpaid volunteers. Clear, observable, and attainable goals can be set (like for a fund raiser) to maintain focus and enthusiasm. Numerous nonfinancial rewards and peer recognition can be utilized. Public recognition, compliments and verbal thanks, letters of commendation, T-shirts, and mementos are all techniques utilized.

8.4.2.6.   The Surgical Team Structure

The surgical team structure is conceptually the opposite of the egoless team. The focus of the surgical team is upon one individual, the surgeon. The surgeon is the possessor of the critical skills around which the project revolves. Everyone on the team provides support to the surgeon. (See Figure 8.6.)

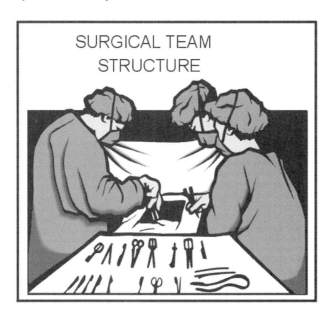

SURGICAL TEAM
STRUCTURE

FIGURE 8.6

In an organizational setting, the surgical team structure would be used when one person is clearly the expert, a scarce resource, or holder of critical skills. For example, there could be a design engineer or scientist with skills in a strategically critical area. A new product design effort might be built around the input and needs of the scarce engineer or scientist.

The surgical team approach was invented by Frank and Lillian Gilbreth in the early 1900s. Frank and Lillian were the creators of time and motion studies, instrumental in laying the foundation of industrial engineering, and seekers of the "one-best" scientific way to accomplish management tasks. Frank Gilbreth started life as a bricklayer apprentice and quickly realized that the key component in the productive process was the journeyman bricklayer. Rather than have each bricklayer mix their own mortar, carry their own bricks, and then lay them, Frank made the bricklayer the focal point of the effort. The journeyman bricklayer was dedicated to only laying bricks and other workers supplied them with bricks and mortar.

Lillian Gilbreth wrote a series of articles about the process for an engineering magazine, and the field of time and motion studies was born. Frank Gilbreth then began studying the application of the organizing principles of bricklaying to surgical teams associated with military battle groups. The military hospitals required rapid decisionmaking and speedy response to save injured soldiers. From all of this was developed the surgical team concept.

Management of the surgical team is strictly authoritarian and autocratic. The surgeon has formal as well as expert authority. Response of team members is immediate and unquestioning.

8.4.2.7. Group Size and Diminishing Returns

The literature indicates that group performance increases only up to a point as the team size expands. At the initial stages of growth new members add cross-functional information and knowledge. As group size increases, efficiency and productivity per individual decreases.

Researchers Wageman and Baker found that higher performance is recorded in smaller groups. Their study concluded that small groups do better at associating performance with rewards. Reward size per individual tends to be larger in small groups. As team size increases, often the total reward remains the same and consequently is diluted among the increased number of team members. Individuals in smaller groups are more committed and have a propensity to set higher goals, find more ways to cooperate, share more information, and spread the workload. Social loafing by team members is less than is found in larger groups. The lesson is that performance can be enhanced by breaking the larger team into smaller teams and subprojects.

> **Standard:** **The best practices project organization organizes projects into small groups, subgroups, and modules.**

References: Karau and Williams, 1993; McGrath, 1984; Shaw, 1981.

8.4.2.8. Team Duration and Its Impact on Performance

Researchers have differing views about the value of group tenure, or the amount of time the group has been working together. Overall it can be concluded that group tenure when combined with group cohesiveness seems to improve performance. Nevertheless, some researchers show that after about three years the effectiveness of the team levels and then declines. It is suggested that groups with high tenure begin isolating themselves from outside critiques and sources of information. The best practices project organization monitors teams for any signs of stagnation. Where needed, fresh team members are added to introduce new thinking and enthusiasm.

There are numerous examples of teams that have worked together for years and maintain high levels of performance. Smith found that patents and technical papers as a measure of group performance consistently increased along with group tenure. O'Keefe, Kernaghan, and Rubenstein concluded that groups with long tenures and high cohesiveness maintained intellectual competitiveness within the group and encouraged innovativeness with resultant high performance ratios. Griffith and Mullins found high levels of communication, organization, and cohesive bonds among research groups that generated revolutionary changes despite resistance from other scientists. They also determined that groups with low tenure lacked development of sufficient role and status relationships.

Porter and his research team found that over time team members become specialized in their particular functional area of expertise, project assignment, and problem areas. As the role differentiation becomes increasingly defined and stabilized, and the perceptions of other team members' capabilities, interests, and contributions are clarified, there is less need to interact within the group. The group knows and expects a predictable output and pattern of behavior from each of the other team members. Efficiency of effort and effectiveness is improved.

As the longevity of the team increases over time, interrelated social influences and processes begin impacting the behavior of the group. In the research, members of groups increasingly shared a common set of beliefs about the project and work setting. Consensus with each other increased until there was a team congruity. Stalk and Hout found that groups tend to organize their work environment to reduce stress and streamline the efficiency, predictability, and clarity of the team's effort determined.

All of these forces can be directed toward maintaining high performance levels, or may encourage *less effective* team behavior. As discussed previously about individual performance, some research suggests that group effectiveness increases to a certain point and then levels off. Several studies even show that many natural tendencies *promote* a performance decline in aging groups.

Shepard's investigation of research and development groups concluded that team performance improves up to 16 months, but then decreases. Pelz and Andrews research, as supported by Smith, determined that the optimum group longevity is between three and four years. The performance decline is encouraged and manifested by several factors.

8.4.2.8.1.   Internal Focus and Isolation.   The group can become increasingly isolated from external forces and sources of information. Researcher Sherif found that groups may become internally focused to the point there is high internal loyalty and strong inner dynamics, but the team is unable to reach out to the external world. As a result the team becomes increasingly isolated from external sources of important information and technology advancements. As time passes the team perceives that there is less of a need to seek audits, critiques, and new information that could impact the team's performance, agreed the findings of Petz and Andrews.

8.4.2.8.2.   Not Invented Here Syndrome.   As group tenure increases with time, there is a tendency to acknowledge only information that is compatible with the team's preconceived viewpoints, interests, and perspectives, determined Rogers and Shoemaker. This is sometimes titled the "not invented here" syndrome. Communications are increasingly targeted to other individuals and groups who reflect the views of the team and are likely to exhibit agreement with the team's approach. Team members even selectively interact with external stakeholders to avoid conflicting messages and information.

8.4.2.8.3.   Group Think.   As teams work together, they develop standardized approaches, work patterns, communications interfaces, and team member relationships that are familiar and comfortable. The shared perceptions that become stronger with time may constrain individual attitudes, opinions, and approaches. Although a great deal of assurance is provided to the team itself, nevertheless, they sometimes become increasingly isolated from outside information, say Petz and Andrews. The research suggests that as a result teams become less and less responsive to unusual challenges and approaches to their task. In particular the team becomes less receptive toward any information that signals a significant disruption of their now imprinted work practices and team behaviors.

8.4.2.8.4.   Egotistical Behavior.   Time combined with team interaction, and natural bonding through cultural and meaningful team experiences, encourage the team to become a homogeneous group. The strengthening cultural ties between team members provide a sense of identity and differentiate the group from others. Researcher Homens found that if the process becomes excessive, it contributes to a sense of superiority over others and isolation from outside communications. Team members increasingly perceive that they possess adequate expertise, knowledge, and skills in their technical areas. It becomes increasingly less important to look to the outside for critiques and new ideas. The team members place their project approach on a psychological pedestal in terms of its perceived technological superiority. They view outside competitive approaches and conflicting approaches as inferior by definition.

References: Ingram, 1998; Toney, 1996; Stalk and Hout, 1990; Katz, 1982; Pfeffer, 1981; Porter et al., 1975; O'Keefe et al., 1975; Griffith and Mullins, 1972; Rogers and Shoemaker, 1971; Smith, 1983; Pelz and Andrews, 1966; Sherif, 1966; Homens, 1961; Shepard, 1956.

8.4.3. Duration Compression

The concept of compressing the duration of the process focuses on reducing the time associated with the critical path. Often the process occurs as the result of a need to speed up a project or to meet a critical time based milestone. From the view of the project organization, encouraging duration compression should be an ongoing process to ensure projects are completed as fast, efficiently, and effectively as possible.

Duration compression is broken into two major approaches: crashing and fast tracking. Crashing is adding resources to the project and fast tracking involves applying superior project manager intellect to the scheduling process.

8.4.3.1. Crashing the Project

The time honored way of speeding up a project (i.e., shortening the duration of the critical path) is by throwing money (or resources) at it. Project managers say that given enough money, they can achieve just about anything. The U.S. Government has repeatedly proven the validity of the concept.

Benchmarkers warn that, "It usually takes a lot of money to achieve a little bit more." For example, when one person is painting a room, adding a person will generally not quite double production. Adding a third person will probably result in an even more diminished amount of improvement. For example, in production there is a heurism; "Never have three people on a job. Two people will work and one will watch." As more and more people are added to the room painting process, each one will contribute incrementally less. At some point, as the room gets crammed with people, total production will even start to diminish.

Even so, benchmarkers advise that when time is the critical element, crashing or adding resources is often the most effective manner in which to solve the problem.

8.4.3.2. Project Fast Tracking

Project fast tracking refers to techniques that accelerate projects while minimizing the need for additional resources. Included in the tool kit of fast tracking techniques is the process of overlapping project phases, the analysis of project schedules in a pragmatic manner, using peer groups to assist in the process, softening project specifications, and transferring activities and involving vendors in the fast tracking process.

8.4.3.2.1. Overlapping Project Phases. Before all the work on one phase of the project is complete, the next phase is initiated. For example, an automotive manufacturer might identify a market need for a new model. Even though the initial marketing study is not complete, if the results appear favorable the company could initiate construction of a prototype. Half way through the prototype evaluation and test, positive results might justify the commencement of work on the preproduction model. In the meantime the initial market studies are completed as is the prototype build and test. Assuming favorable results from the preproduction model, the decision could be made to proceed directly to a pilot run and full production.

Zangwill identified examples of companies that cut lead time to market in half by conducting overlapping product development. Nonaka concluded that a special characteristic of the more successful product introductions that he studied involved overlapping every phase of the development process. Problem solving can also be enhanced by overlapping project phases, disclosed the study by Lorenz. Clark and Fujimoto recorded significant reductions in product development cycle time by overlapping product development phases such as integrating die design and die making. Eisenhardt and Tabrizi recorded a moderately high correlation at $+.38$ between overlapping product development and goal achievement.

References: Zangwill, 1992; Clark and Fujimoto, 1991; Lorenz, 1990; Stalk and Hout, 1990; Eisenhardt and Tabrizi, 1990.

8.4.3.2.2. Pragmatic and Critical Schedule Analysis. Stalk and Hout determined that project speed could be increased simply by critically evaluating the time elements associated with each work package and activity. Often conditions have changed since the original schedule was compiled. In many cases, activities can be shortened without materially affecting the project cost or outcome. Schedule analysis lets the project team eliminate unnecessary steps and sequence activities in a more efficient order.

8.4.3.2.3. Peer Reviews. Gold's research found that the fast tracking process lends itself to a group effort. Teams utilizing peer reviews accelerated project progress.

8.4.3.2.4. Softening Quality, Technical, and Performance Specifications. Often project teams build in higher standards of performance than specified by the customer. A critical "value analysis" approach may result in softer, more easily attainable, and even less costly requirements, according to Gold.

8.4.3.2.5. Involving Outside Suppliers. Projects can be accelerated, often with little increase in overall costs, by buying, licensing, and contracting technology and services from external sources. Supplier involvement reduces project team workload and frees time to focus on the tasks and skills that maximize core competencies and skills. Application of suppliers' skills maximizes synergy. Suppliers also serve as an outside source of critique and elimination of future problems as well as suggestions to improve processes and product.

One word of caution, Eisenhardt and Tabrizi found that early involvement of suppliers seems to be less effective in uncertain projects because often the final suppliers are unknown. In their research the fastest moving project managers tended to delay supplier selection until the final stages of the project.

References: Clark and Fujimoto, 1991; Cordero, 1991; Eisenhardt and Tabrizi, 1990; Stalk and Hout, 1990; Gupta and Wilemon, 1990; Gold, 1987; Kirkpatrick and Locke, 1991; Miller and Toulouse, 1986; Bourgeois, 1980; Weick, 1979; Child, 1977; Lieberson and O'Conner, 1972; Fayol, 1916.

9

Knowledge Management: Providing Structured and Predictable Methodologies and Tools

Benchmarkers say that a major part of the support function of the project organization is knowledge management. Included are areas such as providing predictable and standard methodologies; ensuring that knowledge is transferred from one project to another in the form of a knowledge library; finding, evaluating, and making available the latest tools and software; seeking excellence and setting high standards of performance; and training and teaching of project specific skills for project managers. This chapter addresses methodologies used by the project teams in implementing specific projects. Chapter 10 will cover other knowledge management topics of tools and software provision, seeking excellence, and training.

A major role of the project organization in its efforts to support the project teams is the development and implementation of structured and predictable methodologies, templates, and tools. Included are the best practices, steps taken, the processes followed, and all the methodology employed for project selection and initiation, planning, execution and control, and termination. Project specific standards and best practices instill predictability in the profession by using a common approach that provides a frame of reference for all users in all geographical and cultural environments and can be used by project managers and their management to define and measure expected behavior and performance.

> **Standard:** **The best practices project organization develops and applies a predictable methodology to manage projects.**

Involvement and oversight by the project organization is necessary because the use of a structured approach to project management is not always the *easiest* way for project teams to execute projects. Benchmarking forum participants report that there is a tendency for project managers to skip fundamental steps and gradually return to ad hoc methods of managing projects. Research conclusions by Hise and team agreed with the benchmarkers' judgment. Even though Hise and team's data showed a positive correlation between performing key project activities and a higher project success rate, they also determined that 76% of the project managers surveyed skipped fundamental process steps.

A role of the project organization is to ensure that aspects of methodology that favorably impact goal achievement are followed. Benchmarkers report that project managers have a tendency to eliminate steps because they feel it will save time. When there is pressure to conclude the project the tendency is to bypass or ignore the steps of the process. Resultant effects are weak designs, malfunctions, liability suits, product recalls, and potentially higher production costs of new products being developed. Hise and team conclude that a piecemeal approach to reducing project cycle time does not lead to a successful transformation—only frustration and disappointment rate. Better performing teams manage to be on a fast track and still maintain project quality by performing the fundamental process steps.

The project organization is almost always charged with developing project specific best practices that represent a structured and predictable approach. The approach is methodical yet broad based and flexible. It contains best practices that are appropriate for the broad variety of large projects encountered globally. The most simple methodology or processes appropriate to the project are applied. Core documentation is used on all large projects. The documentation templates can be expanded and detailed to fit the specific project. In the Appendix to this volume are numerous sample templates.

9.1. PROJECT PHASES

Application specifics in the form of best practices and standards are most easily discussed and visualized relative to project life phases and documentation flow. It is stressed that project life phases are to be used to simplify, categorize, and portray the processes, best practices, and standards rather than to be approached as rigid, sequential processes.

Project initiation and selection embrace the techniques of portfolio management and were discussed in detail in Part II. The initiation effort results in the development of the project charter. From this, the project is evaluated and selected by ranking it related to other projects according to strategic factors, financial return and amount of risk. After the project is selected the planning phase includes preparation of the project plan. Included are the details from the project charter as well as an expanded project schedule, budget and risk evaluation and other supporting documents appropriate for the project. During project implementation and control, the project plan is monitored to ascertain project status and variances from the project plan. Scope changes are managed as a formal process. Termination of the project includes a project checklist and communication of project benefits to others in the organization. (See Figure 9.1.)

9.2. FLEXIBLE AND SIMPLE APPROACH

The judgment of the forum is that the broad variety of projects necessitates flexibility in the management approach, and that the superior project organization encourages the project teams to apply the simplest methodology necessary to achieve project goals. Flexibility and speed are key elements of superior project efforts. Although needed are appropriately detailed plans, careful timing, and risk evaluation, the project organization must also face the reality that fast moving, complex, ever-changing scenarios require flexibility and hands-on leadership.

Standard: The best practices project organization uses a methodology that is broad and flexible.

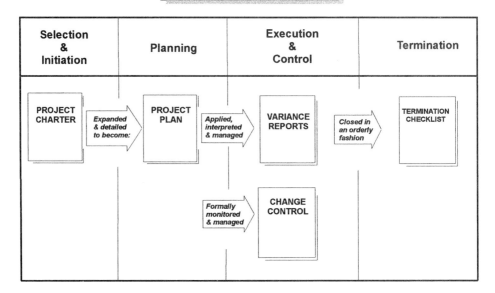

FIGURE 9.1

The importance of applying the least complicated methodology also applies to the organization's inventory of project management tools. Included are scheduling, communications and project specific software, templates, and standardized procedures. The tendency is for the inventory to grow. As the tools inventory grows, it becomes a larger and larger maintenance burden. Benchmarking participants suggest that a simple, standardized set of tools should be used across the organization. Standardization minimizes overlap of tools and reduces training time. New people to the project come up to speed faster. Use of third party vendors and off-the-shelf software is recommended.

9.3. STANDARDIZED METHODOLOGY ADVANTAGES

The employment of a standardized set of project specific best practices and methodology enhances project speed, efficiency, and effectiveness. In addition, it makes the management approach to projects predictable, increases stakeholder confidence, and paints an image of professionalism.

9.3.1. Speed

The structured approach to managing projects (as compared to a random or ad hoc approach) historically has been found to be the fastest way to achieve goals, particularly on large projects as addressed by the best practices in this chapter. Cordero's research determined that exercising a predictable approach to project management tended to simplify the team effort and increased project speed. A predictable and structured approach reduces training time. Inexperienced project managers became proficient faster than if they learned the mechanics of team leadership in a random manner. In a sense, use of a structured methodology is a cookbook approach. The cookbook doesn't necessarily make a person a superior cook, but it does help to get a decent meal on the table in the shortest amount of time.

9.3.2. Efficiency

Project teams are more efficient when utilizing a repeatable and predictable approach. Team members don't need to reinvent the methodology wheel every time they lead and manage a project. The best practices, standards, and documentation for each project phase are easy to understand and implement. They provide a common terminology that results in faster and more accurate communication. The common frame of reference ensures that projects and project management performance can be evaluated, compared, and contrasted between dissimilar projects in various geographical, technical, and organizational environments.

9.3.3. Effectiveness

Increased is the probability of goal achievement. The procedural approach ensures that all steps are taken to attain project success. In cases where projects are marginal or fail to live up to their initial expectations, the best practices result is that they will be promptly evaluated and either fixed or terminated.

9.3.4. Other Benefits

In addition to improvements in speed, efficiency, and effectiveness, the methodological approach to project management is evidenced by other benefits:

- **Improved project management predictability.** A predictable and structured methodology makes the management of projects more predictable. It provides certainty that the large variety of projects even within the same organization will be approached in a similar manner. If the manager of the project is replaced, the new project manager understands the process and can quickly adapt to the project specifics.
- **Stakeholder confidence.** Stakeholders are given confidence about project leadership and the manner in which the project will be run.
- **Professionalism.** It presents a professional image. It lends itself to structured presentations and proposals. It aids managers in easily presenting the concepts and status of the organization to stakeholders.

9.4. STRUCTURED METHODOLOGY CAVEATS

It is stressed that the phases and document flow described in this chapter are presented primarily to simplify and make the approach to managing projects easier to understand. Benchmarking forum attendees report that there is danger in overemphasizing the structured or methodological approach to managing projects. Sometimes a project management methodology is perceived by the organization as a silver bullet that will eliminate all multifunctional project problems. Encouraged is a checklist mentality that believes if all the boxes are checked off, the project will be successful. Benchmarkers agree that such is not the case. The methodology is simply a set of tools for the project manager to use. Success or failure of the project depends upon people performance and a myriad of political, cultural and environmental issues.

9.4.1. Rigid Methodologies May Reduce Project Speed

The tendency of many organizations is to add increasing amounts of mandated methods, templates, procedures and required tool usage. Several best practices organizations said that their methodology had become so detailed, rigid, and bureaucratic that it hindered the management of the project. Benchmarkers are also quick to remind that time spent on methodologies, templates, and tools is time spent away from running the project.

It is generally agreed that over-reliance on rigid methodologies including phase gates and go/no go decision points can reduce the speed of any project, but is particularly harmful for multifunctional projects, projects with overlapping phases, and large projects with numerous subprojects. There is also danger in even categorizing projects by the phases of initiation and selection, planning, execution and control, and termination. The temptation reported by benchmarking participants is to convert the phases to sequential phase gates and go/no go decision points.

The negative impact of the resultant rigidity is particularly evident when using speed based approaches to structuring and managing projects. For example, *overlapping project and design phases* is utilized to reduce total time to project completion. The approach was pioneered in companies striving to accelerate new product development process speed. The concept calls for commencing each new phase before the last is concluded. Conforming to rigid project phases can also reduce the speed and efficiency of *multifunctional teams*. When employing the multifunctional team approach, each functional element often progresses at a different rate of speed. Requiring the faster functional elements to wait for the others to catch up at each project phase is counterproductive. On huge projects with hundreds of subprojects it is improbable that all will be progressing at the same pace or even in the same project phase.

In each of these cases specific project phases are less clearly identifiable. The vagueness of phase gates on large multifunctional projects with overlapping phases and varying subproject speeds makes the task of the project manager more difficult. Benchmarking participants relate the management process requires more constant monitoring and attention. The methodologies are more abstract and lack the strictness of the sequential project life phase approach. More emphasis is placed on subproject milestones and deliverables. The key is to orchestrate all the components to come together at the termination of the project. One company commented that they make a distinction between emphasis on the *process* and on the *management* of the project. After struggling to determine the degree of detail needed, most superior project organizations evolve to using a few fundamental methodology "guidelines" and best practices rather than rigid sets of procedures and templates.

9.4.2. Emphasize Project and People Management

Benchmarkers warn that sometimes software and tools usage are emphasized to the point that management of the project suffers. The consensus of participants is that during the execution and control phase, that project managers should be encouraged to stress face-to-face communications in preference to spending time using scheduling and software tools.

The balance between tools usage and people management is an important consideration for project organizations. Benchmarkers say that often project managers new to the profession suffer from "software addiction." Cases were mentioned during the forums where projects failed because the project manager was fixated on the computer and the project scheduling software and didn't know what was actually happening on the project.

> **Standard:** **Best practices project management methodology emphasizes project and people management over tools.**

As discussed in the portfolio management portion of this book, often the problem is compounded because the *project organization* emphasizes reports and schedule updates. These "demands" for updated information can intrude upon project speed and encourage a project manager mindset that the maintenance of tools, templates, and reports are more important than running the project through face-to-face management of people.

In many cases, project managers simply behave in the manner in which they were trained. Project management courses often emphasize tools usage and an oversimplified checklist approach to project management. Rarely is the same amount of class time spent on managing people and cultural and political problems even though they are often more instrumental in project failure than is the incorrect application of tools. Not surprisingly, surveys of benchmark participants show that generally people new to project management spend up to 75% of their time using the tools of project management.

This is not necessarily all bad according to the benchmarkers. They say that tools proficiency is necessary to form the foundation upon which to build a professional project management approach. As aspiring project managers become increasingly familiar and comfortable with the tools, benchmarkers report that they make a natural and gradual transition to emphasizing people management with as much as 75% of their time spent on face-to-face management.

9.4.3. Progressive Project Phases Management

Project phases become less clear when projects are bid in progressive stages. Progressive bidding occurs when organizations are planning or conducting projects with rapidly changing technologies, the customer wants the latest technology incorporated as the project is progressing, and/or when the project approach to future stages is unknown, vague, or evolving. Both the bidding company and the customer face the dilemma that for other than the current stage, the manner in which future stages will be performed is uncertain. In many such situations the approach is to commit to a fixed price for the immediate stage and for the bidder to give a definitive estimate for the next stage and rougher estimates for stages farther in the future. In a sense, each successive phase then becomes a subproject with its own series of miniphases and activities. Note that the process requires a high degree of trust on the part of the customer. Benchmarking participants using the approach report that customers like the process because it removes much of the uncertainty related to the financial impact of scope changes. From the viewpoint of the bidder it makes cost estimating more critical because profit must be built into the project rather than expecting it from scope changes.

References: For a detailed discussion about the mechanics of each phase of project management, see *A Guide to the Project Management Body of Knowledge,* published by the Project Management Institute, and *Project Management: A Managerial Approach,* by Jack R. Meredith and Samuel J. Mantel, published by John Wiley and Sons. See also Hise et al., 1989.

9.5. PROJECT PHASES AND ASSOCIATED BEST PRACTICES

With the preceding cautionary statements to guide the way, benchmarking participants agree that there are a few *fundamental standards, best practices, and documents* that optimize the probability of goal achievement on nearly all large projects. They are broad and flexible in application, yet provide a predictable approach. Following is a discussion of these standards, best practices, and documentation as they apply to each of the phases or process stages of a typical large project.

9.5.1. Project Initiation and Selection

As was discussed in Part II, project initiation and selection includes all the activities associated with generating a new project concept and then ranking its organizational value related to the portfolio of existing and other proposed projects. From this process the project is selected (or rejected) and initiated. The primary document in the project initiation and selection phase is the project charter.

9.5.1.1. Project Idea Germination

The best practices project host organization has a process for ensuring that all significant business opportunities are evaluated. The process recognizes new projects generated from such sources as response to customer and market demand, internal business needs (i.e., training, services for other departments), research based concepts, and legal and mandated projects (i.e., the government or other source require it).

For projects generated from within an organization, a point stressed by benchmarking participants is that the process must be open to all people in the organization that desire to propose new ideas and projects. Once the idea is proposed a formal process ensures that it is reviewed in an arm's length, independent, and objective manner.

9.5.1.2. Defining the Project Need

Whether the project is internally or externally generated, benchmarking forum participants stress the importance of clearly defining the project need and requirements. In some cases, particularly when bidding projects external to the organization this part of the process can be quite detailed.

The benchmarking participants report that there is a tendency for the quotation for project services to lack specifics and occasionally for promises to be made that the project team can't keep. One participant stated that if there is sufficient information available to develop a price quote, there should be sufficient and clear enough information to develop a set of detailed project specifications. Another made the point that once the project is under way, it is too late to define project need and requirements.

Standard: **The best practices project methodology includes a process for thoroughly defining customer need.**

There is general agreement from benchmark participants that the ultimate measure of project performance is customer satisfaction.

At the initiation stage of the process, the satisfaction building effort starts by motivating and assisting the client or stakeholder to define and specifically identify what is wanted from a project. Emphasis is placed on having the customer involved on the project development team. Assumptions should be identified early on and a determination made about where changes will likely occur.

Some benchmarking participants work with customers to help them write the project needs and requirements. They have experienced facilitators whose role is to assist customers in writing the details of the project to be performed. After the rough details are compiled with the customer, a meeting is held with technical experts to refine and finalize the details. Another company sent their facilitators through a "requirements writing" course. The facilitators tend to be nontechnical with a broad base of knowledge about the business, legal, and performance aspects of the proposal. Technical people are used in an advisory role during the rough draft phase.

9.5.1.3. Project Selection

When project management is involved with portfolio management and selection, organizational financial return is increased because a pragmatic decision process ensures that each project serves to optimize the overall portfolio financial and strategic value. In addition, the management of specific projects is more focused because the project team is aware of the relative importance of each project and how it helps achieve the company's strategic objectives.

To achieve the benefits, the project selection activities must be performed with professionalism. Dwyer and Mellor studied new product development projects and found a clear positive relationship between proficiency in analyzing and selecting projects and ultimate new product success. In companies with the highest records of new product success, they found superior up-front and predevelopment activities, initial screening, extensive preliminary market assessments, thorough business and financial analysis, and preliminary technical assessment.

Similar research by Cooper determined that superior project results were associated with comprehensive evaluation and analysis of pertinent success factors such as market demand and competition, understanding of project activities and deliverables; clear definition of the customer's needs, wants, and behavior; definition of buyer price sensitivity; and expected involvement from associated functional areas.

A follow-up study by Cooper in association with Kleinschmidt added detail to Cooper's earlier conclusions. They determined that efforts prior to project initiation had high correlation with subsequent project success. Included was determination of a clear understanding of what the project related product or service would be and do. When the project involved new product development, success was linked to detailed market studies and marketing research. Determination of the fit of the project with the portfolio of projects and corporate areas of expertise was also a key factor. Cooper and Kleinschmidt recorded that the most successful project groups had a high degree of technological synergy between the project's product or service and the host organization. The selection team ensured that there was a good fit between the needs of the project and the host organization's skills and resources.

9.5.1.4. The Project Charter

The document that confers rights and authority to the project team is the project charter. Its antecedents are ancient. Its use dates at least to the Magna Carta ("The Great Charter") of 1215, a document that continues to serve as a constitutional cornerstone of the United Kingdom.

The project charter is the key document associated with the initiation and selection phase. It provides information needed to make a decision about the feasibility of the project and authorizes further action. It is the project based counterpart to the business plan in organizational strategic management. It serves as the primary tool used to compare one project with another when analyzing and optimizing the project portfolio. Once the project is accepted, signature approval of the project charter launches the planning phase of the project. The project charter then becomes the foundation or starting point for development of the more detailed project plan.

Shown in the sidebar is a summary of a generic project charter. For more information the appendix contains a detailed project charter checklist. Note that the specifics addressed by the project charter information vary from organization to organization and project to project.

The project charter gives background information about the project that is pertinent to the decisionmaking process. It also reviews the need that generated the project request and discusses the project goals and objectives related to the host organization's strategy and goals.

The project charter evaluates the project based upon estimated financial return, risk related to other projects in the portfolio, and subjective factors. The analytical discipline is the same as found in the management of any portfolio, of valuable assets (stocks, bonds,

Project Charter
Project Selection Decisionmaking Information

Background
  Need for the project
  Alternatives investigated
  Compatibility with the host organization's strategic plan
  Project goals and strategies
Selection and Evaluation: ranking criteria
  Financial return and portfolio ranking: net present value, internal rate of return, payback
  Risk analysis: risk related to the portfolio of projects
  Subjective factors: fit with existing services or product lines, constraints, assumptions, market & strategic positioning, public image
Scope: overview and details pertinent to the decision process
  Initial work breakdown structure
  Deliverables
  Functional requirements
  Technical requirements
  Time schedule for project completion
  Budget and cash flow
  Risk analysis (project specific)
  Quality plan
  Communications plan
  Key success factors
  Performance measurement metrics
  Scope management and control: project control, variance reporting structure and procedures, measurement metrics
  Information systems
Supporting Documents
Signature Page

---

properties, etc.). Each project is evaluated related to the overall portfolio of projects and organizational objectives.

Defined is the scope of the project, which in a simplified sense is a description of the activities and deliverables that involve money and other resources. The project charter presents a graphical overview of project activities and deliverables in the form of a work breakdown structure. Also included are appropriate levels of detail regarding the schedule, budget, project specific risks, quality targets, and important stakeholder communications issues. Depending upon the degree of detail considered important to the decisionmaking process, the project charter discusses project control in the form of variance analysis, scope management, and available information systems. The project charter is a formal document. Any project where more than incidental expenditures of resources are budgeted demands a signature to authorize and activate the project planning phase.

> **Standard:** **The best practices project methodology includes a project charter that contains all material information needed to make a decision whether to approve the project.**

9.5.1.4.1. Project Selection and Ranking Criteria. The project charter will normally contain a ranking of the project under consideration compared to other projects in the portfolio. Projects are ranked because there are rarely if ever enough resources to accept every favorable project. The objective of the selection process is to maximize the strategic effectiveness and financial return of the *portfolio of projects*, while minimizing portfolio risk and optimizing subjective factors such as the attainment of organizational goals.

> **Standard:** **The best practices project methodology selects projects by evaluating and ranking strategic factors, financial benefits, and risk.**

To attain the objectives, each prospective project is analyzed and ranked with other opportunities according to strategic factors, financial return, and risk (See Part II for a detailed discussion of selection and ranking criteria in detail.)

- **Subjective factors.** All the factors that impact the project's viability but can't be quantified are labeled subjective. Included might be considerations such as, "Does the project fit our overall organizational strategy?" "Is it strategically critical?" "Is the project needed to fill in gaps in the product line?"

Benchmarking participants report that often subjective factors are more important to the decision to initiate work on the project than are financial and risk related variables. For example, one individual managed an asset portfolio for a large conglomerate consisting of 80 different companies. The individual's job was to determine which companies to sell and which new companies to purchase. When asked which types of tools were used to make the determination, he responded that all the conventional ranking methods were used including NPV, IRR, payback, and asset return calculations. However, he also stressed that often *none* of the mathematical methods were used. He said that the companies within the portfolio that were extremely bad and extremely good were so obvious that little ranking analysis was needed to support a decision. To justify the logic, he used an example in a person's personal life. If the individual were driving home from work and the transmission went out in their car they would intuitively know without the benefit of analysis that an investment was immediately required to fix the transmission. In the same decisionmaking vein, employees in large organizations are often offered opportunities to purchase company stock at greatly discounted prices. It takes little or no analysis to determine that it is an attractive offer.

Included in the analysis of subjective factors are project constraints. Constraints may be arbitrary in nature. For example, many new product development projects work under a "no new technology" constraint. The constraint reduces scope changes and emphasizes reaching the market in the shortest time. Often the no-new-technology constraint is a decision that must be made by senior management.

- **Financial return.** People tend to intuitively rank investment opportunities by how much money each opportunity is expected to earn. The validity of the intuition is well founded. Numerous researchers have shown a clear positive correlation between the financial analysis of project opportunities and project goal achievement. Financial return analysis involves calculating how much money (i.e., revenue, cash flow) the project will generate compared to its cost. Techniques used commonly involve the calculation of net present value (NPV), internal rate of return (IRR), and payback period. From these calculations, projects can be ranked according to those expected to generate the most revenue down to those expected to generate the least.

As shown in Figure 9.2, project selection by analyzing potential financial returns is the first step in a series of financial activities and documentation that flows from beginning to the end of the project. Before the potential financial return for the project can be calculated, most project organizations put together an initial budget and cash flow projection. The cash flow projection shows all cash coming into the project as well as flowing out. It can be calculated for any period of time deemed suitable but usually is monthly for the near term and yearly for periods farther into the future. From the cash flow or tentative budget can be calculated net present values, internal rates of return, and payback periods for the project. Some companies also use a variety of other financial measures such as return on equity and benefit cost ratios.

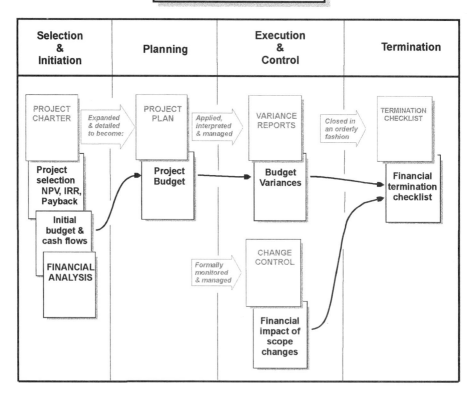

FIGURE 9.2

After the project is selected and approved, the planning phase includes the development of the detailed budget. The detailed budget contains an appropriate level of detail for the project at hand, often including estimated cost for each work package in the project plan.

When the project is underway and being executed, the estimated costs and revenues are compared with actuals. The variances are periodically reviewed and analyzed. Project control is maintained through an appropriate response to abnormal variances.

Any changes to the scope of the project could affect project cost. The impact of the change is calculated. Project baselines are adjusted and customers are charged for the change.

During the project termination phase, a financial termination checklist ensures that all financial activities and documentation are addressed.

- **Risk analysis.** The superior project manager identifies, analyzes, and manages risk throughout the project. As shown in Figure 9.3, the selection and initiation stage of project management involves two fundamental types of risk analysis. The first is *portfolio* risk. It involves the overall risk associated with one project when compared to another. The second is the initial analysis of risk *specific to the project itself.*

  The planning phase of the project involves detailed risk analysis of the specific project. It includes identification of each potential risk event, its estimated probability of occurrence, potential financial consequences as a result of the risk occurrence, and planned response, if any, to each risk.

  The execution and control phases find the project manager monitoring the project for the development of risk events and then adjusting accordingly. At project termination any significant risk events are analyzed and their impact on the project is determined. Recommendations are made to help future project managers develop appropriate risk responses.

9.5.1.4.2. Scope. The scope of the project is discussed in the project charter from the viewpoint of its relationship to the initiation and selection decisionmaking process. Generally included is an overview of the process and discussion of salient details in pertinent areas listed in the following.

- **Initial work breakdown structure.** A graphical depiction of the major activities and deliverables of the project can be

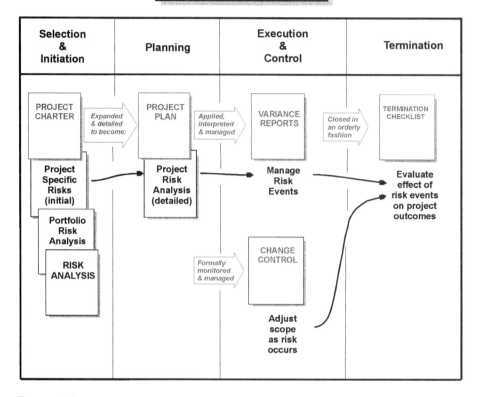

RISK ANALYSIS DOCUMENT FLOW

Selection & Initiation	Planning	Execution & Control	Termination

FIGURE 9.3

presented. The objective of the initial work breakdown structure is to help decisionmakers visualize the project. It is an
effective way to show the size and complexity of the project.
It also serves as the platform to discuss other elements of the
project affecting the decision process such as the schedule,
risks, quality, and communications plans.

- **Deliverables.** The specific services and products that are to
be delivered to the customer focus the selection discussion.
Included could be subdeliverables that the customer might
receive as the project is progressing. Some benchmarking organizations working with intangibles such as software, have
guidelines that the customer is delivered some form of visible
outcome at each milestone or even periodically such as each
six weeks. Deliverables could also include subdeliverables if
the project has many subprojects. For example, if the project

were the construction of a prototype aircraft, there would be numerous subdeliverables for the cockpit instrumentation project.

- **Key success factors.** The project charter will likely discuss the key factors upon which the customer and other stakeholders will judge project success.

- **Performance measurement metrics.** After establishment of the key success factors, the project charter may contain details about the manner in which the success will be measured. The performance measurement metrics are clear, easy to understand, observable, and measurable. Part II contains a detailed discussion of types of performance measurements.

- **Functional specifications.** The expectations of the customer are described by the functional specifications of the product or service. The functional specifications describe what the product or service will do. For example, if the project were the development of a new telephone, the functional specifications would describe the features such as wireless capability, call waiting, and the other features desired by the customer.

- **Technical specifications.** The manner in which the functional specifications will be achieved is described by the technical specifications. If the project is developing a product the technical specifications describe the approach to constructing the product. In the case of the telephone, the technical specifications would describe the design of the telephone.

- **Schedule.** For selection and initiation generally the time schedule portion of the project charter includes critical dates, constraints, milestones, go/no go decision points and delivery dates.

- **Budget and cash flow.** To have adequate information to develop the net present value and internal rate of return calculations, it is necessary to prepare a cash flow forecast. This cash flow forecast will become the foundation of the detailed budget that is prepared in the planning phase of the project. The purpose of the cash flow projection is to show the revenue and expense streams over the life of the project. Generally the initial cash flow analysis is less detailed than the budget developed during the planning phase of the project. It is typically cash flow based with few if any accrual accounting related modifications to the cash flows. This information can be used through the present value process to compare the prospective project with others being considered.

- **Quality plan.** Any pertinent details, targets, constraints or customer requirements regarding the quality of the project service or product are introduced in the selection and initiation phase.
- **Communications plan.** Critical stakeholder communications variables are introduced. Communications factors that affect the selection decision are discussed.
- **Scope management and control.** Included are project control, variance reporting structure and procedures, and measurement metrics. Particularly on projects where scope change represents additional revenue and where scope changes represent serious financial consequences, the project charter will discuss scope management and control.
- **Information systems.** New project tools are introduced periodically and existing tools are evolving constantly. Where pertinent to the decisionmaking process, the project charter will discuss the tools to be used and their expected impact on project goal achievement.

References: Dwyer and Mellor, 1991; Gupta and Wilemon, 1990; Cooper and Kleinschmidt, 1987; Cooper, 1979; Rothwell et.al., 1974.

9.5.2. The Planning Phase

Planning defines the approach taken to execute the project. It is the expansion and detailing of the project charter into the project plan. The project plan includes a definition of scope with emphasis placed on development of the project schedule and critical path, detailing of the budget and analysis of project specific risk. Other activities included in the project plan include appropriate amounts of quality planning, stakeholder communications planning, and staff and resources scheduling.

A study by Cooper and Kleinschmidt determined that planning had high correlation with subsequent project success. Included should be a definition of the customer's needs, wants, and preferences; determination of a clear understanding of what the project related product or service will be and do; and detailing of the project's product or service functional and technical specifications and requirements. The research team found that the steps that occur before a project gets underway are critical and that management and the project team should be prepared to devote time, resources, and money to ensure that the planning phase is accomplished in a superior manner.

Researcher Pfeffer found that development teams that spent more time planning had an edge in gaining resources. Maidique and Zirger determined that good planning has a positive correlation with project goal achievement. Extensive planning serves as a signal of project quality, judged Feldman and team. Eisenhardt and Tabrizi determined that detailed planning had a positive correlation with project success. More complete planning lets project developers better understand and refine the process, said Gupta and Wilemon.

Note that Eisenhard also found that in fast moving projects with rapidly changing technologies that excessive planning slows down the project. Excessive planning added rigidity and inability to adjust to changing scenarios, alternatives, and opportunities. In those cases, the key element was the superior project manager who could manage the many variables and plan the near future as the project unfolded.

Although planning should be conducted to an appropriate degree, Zangwill found that the natural tendency was to insufficiently plan. Speed based project teams and management focus discouraged time-consuming planning. Zangwill observed situations where extensive planning was constrained by the accounting people and budget restrictions. Other organizations removed personnel from projects that were just starting and assigned them to projects behind schedule. The logic was that the people weren't needed as badly on the new projects. The process fed upon itself because a major reason the other projects were behind schedule was because they were insufficiently planned. Zangwill concluded that it is better to add five planners early in the process than to add 50 people to solve problems farther down stream.

9.5.2.1. The Project Plan

After the project charter is evaluated and approved, attention shifts to the development of the project plan. It may be thought of as an expanded and detailed version of the project charter. It is the fundamental document that guides the project and serves as the mechanism to establish control and measure performance. It is the basic reference document that defines core topics as deliverables and milestones for performance evaluation.

As summarized in the side bar, most project plans are comprised of sections dealing with the background of the project, a detailed description of the scope, management, and control of the project during the execution phase, and the manner in which scope changes will be managed.

The background section of the project plan is generally taken from the project charter and often is unchanged. It typically includes

The Project Plan
Information Needed to Achieve the Project Goals

1. **Background:** pertinent items from the project charter
   Strategy and goals
   Need
   Other: assumptions, unresolved issues, etc.
2. **Project Scope:** more detailed than the project charter
   Work breakdown structure
   Deliverables
   Critical success factors
   Schedule
   Budget
   Risk analysis
   Functional specifications
   Technical specifications
   Quality plan
   Communications plan
3. **Management and Control**
   Measurement metrics
   Variance reports
   Audits
   Milestones, decision points
4. **Scope Change Management**
5. **Signatures**

discussion of the project strategy and its relationship to organizational strategy as well as a definition of the project need, deliverables, and other pertinent assumptions and issues. Background information is an important element of the project plan because it focuses the scope and control sections on the project and organizational goals.

The project scope section details all of the areas that involve the expenditure of money and other resources. On nearly all large projects the scope section will cover, as a minimum, the work breakdown structure, a schedule, budget, project specific risk analysis, and the functional and technical specifications. Depending upon the project and information considered appropriate the scope section could include discussion about resource allocations, the quality plan, and the communications plan.

The management and control section of the project plan discusses the execution phase of the project. It details critical success factors the customer will use to judge the project. Variance reports detail the methods of measuring and controlling project metrics such as the schedule and cost tracking. For most projects audits are scheduled to provide an independent and objective evaluation of project progress. The execution phase will define various milestone and other decision points and the material to be reviewed and action taken at each step of the way.

The final area of the project plan is the signature page. All details of the project plan should be agreed upon and signed off before proceeding with the project. Once signed, the specifications are frozen and any changes can be made only via a formalized scope change control process. Freezing the specifications has the added advantage that it forces people to more adequately plan the project.

9.5.2.2. Work Breakdown Structure

The work breakdown structure reduces project complexity and nebulous goals into a graphical outline or picture of the project. It is a reflection of the common understanding of the team and stakeholders about the scope of the project. It serves as the starting point for other project activities including the development of the project schedule, budget, risk analysis, communications, and quality plan. It is used to show other people the overall view of the project. Benchmarking forum participants suggest that the work breakdown structure is best presented in a visual format. Although it can be shown as a listing of major deliverables and/or activities, the visual format is easier to view and understand by users and stakeholders.

Standard: **The best practices project methodology requires a graphical work breakdown structure for the project.**

Benchmarking forum participants report that people new to the profession often skip the work breakdown structure. The temptation is to proceed directly to the project scheduling software and begin entering activities. When the process is ended the individual has numerous pages of lines depicting the Gantt chart. When asked to describe the project, the only way it can be shown or discussed is to present the multipage listing of the Gantt chart. Benchmarkers agree that a better way is to show the project in the form of a work breakdown structure.

The work breakdown structure shows the major project elements decomposed at each descending level into smaller components. The lowest element is the work package. In common usage the work package is the smallest element that could be contracted out. It usually is less than 80 hours duration.

For example, the highly simplified Figure 9.4 depicts a home building project. It consists of preconstruction activities such as selecting the site, model, and builder. Finances are a separate sub project. The home building schedule involves construction of the foundation, frame, roof, and a myriad of associated activities. Shown in a box under the foundation is the work package consisting of having people dig the footings for the foundation. A common numbering system is used in the example to serve as a reference.

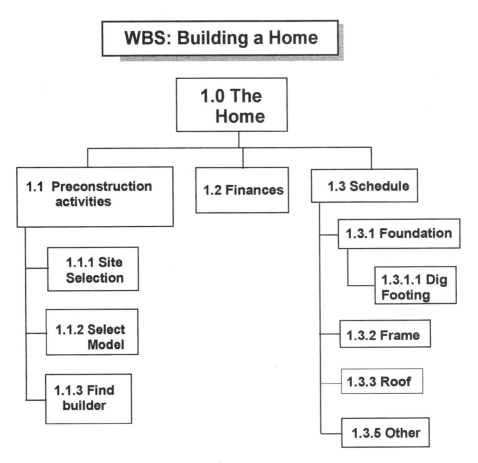

FIGURE 9.4

References: Pfeffer, 1992; Zangwill, 1992; Maidique and Zirger, 1990; Gupta and Wilemon, 1990; Eisenhardt, 1989; Cooper, and Kleinschmidt, 1987; Feldman and March, 1981a.

9.5.2.3. Deliverables

The project plan details the specific services or products that are delivered to the customer. In the case of building a house, the final deliverable would be the house itself. Many projects have numerous subdeliverables. Software projects might include the delivery of various visual products as the software is developed. The deliverables represent a tangible and visible component that the customer uses to judge project success. On the house building project the subdeliverables could include the delivery of the foundation, the framing, the roof, etc.

9.5.2.4. Critical Success Factors

Best practices project managers define the factors against which the performance of the project team will be evaluated. Usually the focus is on customer requirements, but other stakeholders could have significant influence on the perception of project success or failure as well. To determine critical success factors the project manager might ask, "What are the standards that we will be measured against?" Often an organization will have performance records from similar projects executed in prior years. Once the critical success factors are determined the next step is to define the measurement metrics that will be used to measure project performance.

9.5.2.5. The Schedule

The schedule for the project is completed in an appropriate level of detail. Normally shown in the project schedule are sequences of activities, durations, and the critical path. Typically the work breakdown structure serves as the foundation upon which the project schedule is derived. All benchmarking organizations use off-the-shelf software packages to list schedule activities and visually depict the schedule in the format of a Gantt chart, PERT diagram, or flow diagram.

Standard: **The best practices methodology includes a schedule showing all appropriate activities, sequences, durations, and the critical path.**

9.5.2.5.1. Resource Plan. A projection of the use of resources is generally a part of the project schedule. Critical resources
are usually cash and people, although other factors such as specialized machines and services are common. Usually project managers
compile the schedule and budget appropriate costs and resource usage for each work package. The result is a clear picture of the under-
or over application of resources over the expected life of the project.
These resources are then adjusted to fit the constraints and timing
of specific project activities.

9.5.2.5.2. Milestone Emphasis. Every organization in the
benchmarking forum emphasizes the use of milestones to define targets and provide schedule control. Milestones set specific dates for
completion of activities or project phases. On complex projects with
multifunctional approaches, overlapping design phases, and long
spans of time between phases, milestones are applied to subprojects
and the activities of individual areas of action. Often a milestone is
accompanied by a review or audit.

Eisenhardt and Tabrizi's investigation determined that highest
performing project teams used milestones as measures of project
time, cost, and scope performance. They found that milestones have
the intangible benefit of giving the perception of order and routine.

Milestones can also increase project speed. The conclusion of researcher Gersick was that frequent milestones reduce project cycle
time because they are motivating. Their immediacy creates a sense
of urgency and deadline that discourages procrastination. Frequent
milestones were found to accelerate project processes because they
forced people to review the project often. If the project was found to
be off course, it could be corrected early in the process. Milestones
compel performance reviews at periodic intervals and are easy to use
and understand. Rewarding project teams for achieving clear deadlines and milestones synchronizes team energies. Gersick also found
that milestones are particularly important in unstable situations.

The use of milestones for time management, control, and performance measurement is clearly preferred by benchmarkers over using
earned value calculations. Of approximately 60 companies in the
benchmarking forum, all preferred milestones to earned value. The
major reason given was that earned value uses *dollars to measure
time* and milestones use *time to measure time*. For example, consider
a couple building a home with a delivery date, or milestone, of December 1. On December 5th, using the milestone concept, the house delivery is five days behind schedule. If the contractor followed the earned

value approach, when asked about the lateness they might respond with statements such as, "I am only working at 80% efficiency" or "At this point I had expected to spend $100,000 but have only spent $80,000, therefore I am behind schedule." Intuitively most project managers prefer using time to measure time rather than money because it is easy to understand and measure. Note that on projects where performance must be evaluated *at a point between milestones,* then earned value is the most appropriate method to be used.

> **Standard:** **The best practices project methodology uses frequent milestones to measure schedule performance.**

References: Eisenhardt and Tabrizi, 1990; Gersick et al., 1988.

9.5.2.6. The Budget

Budgeting and cash flow projections are accounting and financial activities shared by most project managers. The manner and accuracy in which budgeting is performed can have a significant impact on the project. If the budget is biased, it will cloud stakeholder's views of the project leader's integrity as well as the overall knowledge base. For the project organization it can alter the selection of projects. For the host organization, it could impact the bottom line as well as the attainment of organizational goals. Budgeting and cash flow forecasting should reflect the external environment and expected changes if the highest probability of project success is to be attained.

The most common discussion of benchmarking participants related to budgeting concerns the issue of budget padding. When developing a detailed bottom up budget, most project managers estimate the cost of each work package. The tendency is to add a cushion, or to pad, each work package. The end product is a budget that is higher than a nonpadded estimate.

Many organizational cultures tacitly encourage budget padding. Some organizational cultures have negative views of projects that come in over budget and penalize the associated project manager. The usual response is to make sure that every project comes in *under* budget. Budget padding is also prevalent in organizational cultures where the budget is inflexible and slow to adjust to emergencies or rapidly changing conditions.

The generally agreed best practices approach to budgeting is for the superior project manager to develop the most accurate budget estimate possible. A contingency factor or percentage is then added to the overall project to compensate for risk, emergencies, and other unknowns.

As an example, one benchmarking participant uses the contingency fund to ensure that projects never *appear to be* over budget, while at the same providing ability for precise project performance measurement. The project group constructs large complex facilities and judges project performance by measuring accuracy in attaining budget goals. Coming in five percent *over* budget is viewed the same as coming in five percent *under* budget. However, the extenuating circumstance is that the organizational culture is such that coming in over budget is considered to be bad. Consequently, the project group has a "floating contingency factor" that is arbitrarily placed at 15% of the estimated construction cost. If the contingency is not used on one project it carries over to the next. Since half the projects are under budget and half are over, the contingency is never used up. Stakeholders are happy because all projects come in under budget and the project performance evaluators are pleased because they can precisely measure accuracy of performance.

Standard: **The best practices project methodology includes a project budget (or cash flow) that is appropriately accurate for the scope of the project.**

9.5.2.7. Project Specific Risk Analysis

The project initiation and selection phase evaluated the risk of the individual project compared to the portfolio of projects. It also investigated *major* risks associated with the project itself. The project plan looks more thoroughly at the project specific risk. In particular it investigates and details each significant risk event, its likelihood of occurrence, the magnitude of its consequences, and the response plan.

Incorrect estimation of project risk can be devastating. The research of Ingram focused on failed projects and the impact on the host organization. Ingram found that often large functional organizations assume their expertise in their industry will carry over to the management of large projects. They assume success and ignore or minimize the estimation of the risk and associated cost of total project failure. Ingram found that the failure of large publicly reported projects negatively impacted stock prices of the host organization as much as 40%. His conclusion was that the risk of total project failure should be estimated as well as the public relations impact of the failure and effect on other company areas.

9.5.2.7.1. Risk Responsibility. Benchmark participants report that a common problem encountered by project managers is uncertainty about risk ownership or the determination of who is responsible for the negative impact of risk occurrences. The problem is particularly acute on high technology projects where the customer has limited knowledge of the technology.

The general consensus of the benchmarkers is that the *customer* has ultimate responsibility for project related risks. It is the role of the project manager to identify the risks and communicate them to the customer. It is also agreed that it is necessary to clearly specify and define with the customer their ownership of the risk. For example, consider the situation where a couple desires to build a new home on the flood plain of a river. If the river should flood, it is the homeowner's responsibility and not that of the contractor constructing the house. The selected response to the risk is the burden of the customer, or homeowner, not the contractor. It is the home owner who should decide upon a course of action such as mitigating the risk by building a dike around the house, deflecting the risk by obtaining insurance, or avoiding the risk by building on a hill. The role of the best practices project manager is to communicate significant potential risks to the project owner along with their probability of occurrence, magnitude, and suggested response.

9.5.2.7.2. Risk Analysis. Benchmarking forum participants agree that there is a need to improve the analytical approach to risk assessment and evaluation. During the planning stage, effort should be made to define what can go wrong and what the probability and effect of the risk are. If these actions can be identified, quantified, and combined into a predictive model, a response can be developed and the probability of achieving project goals can be measurably improved.

To assist the project manager's efforts, there is a growing research base of specific actions that affect projects and increase the probability of project success and those that increase the risk of failure. In appendix C is a generic risk evaluation template. Every factor on the template can be scientifically correlated with project success or failure. A user of the risk evaluation template would complete the chart and fill in risk categories and checklist items. The end product spotlights high risk categories and items. A risk mitigation, diversion, or minimization plan would be developed for items in the checklist that represent significant risk to the project.

> **Standard:** **The best practices project methodology identifies and analyzes project specific risk events, determines their likelihood of occurrence and significance, and plans appropriate responses.**

The risk analysis evaluation should be a periodic and ongoing process, and reviewed at each project milestone. At each review, the relative risk of various factors should be evaluated and updated.

9.5.2.7.3. Risk Quantification. As discussed in the project initiation and selection section most organizations evaluate project portfolio risk by estimating "high-medium-low." It is also the most popular method for classifying project specific risks. Other risk quantification and evaluation methods are

- **Probability.** It is also possible to apply probabilities of success to various alternatives and subalternatives. An example in the world of project management would be to develop a probability decision tree with each branch of the tree showing the expected probability of occurrence and the expenses or gains associated with the occurrence. The method is particularly appropriate for complex projects with numerous options unfolding at various project phases.
- **Simulation.** Typically simulation analysis refers to the use of spreadsheets to "change one thing at a time" and then to analyze the effects of the change. For example, an income statement or cash flow might be developed for a new product. If the impact of a pending recession were to be evaluated, the team could estimate the impact (say a 20% reduction in sales), plug that into the spreadsheet, and then observe and evaluate the results.

9.5.2.7.4. Risk Reduction and Mitigation. Once risk is identified, it is necessary to make plans to accept, avoid, deflect, or mitigate the risk. A few guidelines are now listed.

- Include several decision points, milestones, or go/no go points.
- Include capability plans to shut down, terminate, spin off, or find alternative uses if the project gets into trouble.
- Keep fixed costs minimal and minimize financial leverage.
- Diversify.
- Evaluate project success at arm's length (e.g., don't get emotionally involved).
- Emphasize fast payback. Pull out seed money early.
- Have an audit plan. Define problems early, remedy them, and avoid them in the future.
- Preidentify major risk factors in the project. Monitor closely.
- Modularize; break the project into independent components or modules. This can take the form of components of the product or service being developed or can even be time modules such as milestone go/no go decision gates. When working on a complex project, the benefit of modularity is that if one element of the project turns sour, the remaining project modules are unaffected.
- Transfer risk to others (i.e., insurance), use outside contractors, have partners to share the risk.

References: Ingram, 1998; Zangwill, 1992.

9.5.2.8. Functional Specifications

The portion of the project plan that describes customer expectations for the product or service is the functional specifications. Using the example of a telephone, the functional specifications would define features such as wireless communication, ability to be used globally, call waiting, voicemail, and internet access. Also included would be specifications about other customer expectations including product life, serviceability, maintenance, durability (resistance to damage), size, weight, colors, and appearance.

9.5.2.9. Technical Specifications

Details about the technical design, construction, and approach of the product or service are identified by the technical specifications. The technical specifications describe *how* the functional specifications will be achieved. For the telephone used as an example in the preceding paragraph, included would be factors such as materials and components expected to be used, circuit board design, national and international standards to be followed, and technology applied.

> **Standard:** **The best practices project methodology prepares an appropriately detailed set of functional and technical specifications for the project product or service.**

9.5.2.10. The Quality Plan

Usually, the technical specifications will include an appropriately detailed quality plan. The plan will describe quality *assurance* elements for which the project team assumes responsibility and quality manages itself. Also covered are quality *control* subject areas that define the manner in which nonproject personnel will audit and measure project quality performance.

9.5.2.11. Stakeholder Communications

The discussion earlier in the book about project manager professionalism emphasized the linkage between effective communications and project goal achievement. The case was made that the ability to communicate an appropriate amount with project stakeholders is a major factor that differentiates the superior project manager from all others. The research is also clear that the communications process is originated, defined, and stimulated by the project manager.

Benchmarking forum attendees agree that a communications plan should always be developed for large projects being conducted in functional organizations. Included are such topics as the type and frequency of communications with each stakeholder group and whether it is formal or informal. The communications plan might also describe conflict resolution procedures and approaches.

> **Standard:** **The best practices project methodology prepares a communications plan that provides appropriate frequency and detail for the needs of each stakeholder group.**

9.5.2.11.1. Political Factors. Development of the communications plan often includes an analysis of the political factors impacting the probability of project success. A tool for analyzing such political issues is the Sayre wheel. Wallace Sayre dedicated himself to the development of a method to graphically show and quantify the factors impacting an issue and its resolution in the U.S. government. He determined that multiple groups impact the process and that it is necessary to separate and define each group, quantify the probability of the group changing its views, and then to plan a strategy for each group. His approach is particularly appropriate for political issues involving many stakeholder points of view. Note that the Sayre wheel might not be a part of the communications plan itself but could be an internal confidential planning tool.

Figure 9.5 shows the Sayre wheel applied to the political aspects of initiating a project office in a large functional organization. The project is shown in the center of the wheel and various stakeholder groups are identified on the ends of the spokes. In this example, the CEO is judged to represent about 30% of the determination whether the project will be successful. Executives and managers of the impacted functional areas are also felt to be key stakeholders who will be working with the project management office and impacting its success. The same approach is followed with other perceived stakeholders. In each case, a judgment is made about the likelihood of changing the views of the stakeholders and the type and amount of communications or actions necessary. By estimating the impact of each group on the expected success of the project it is possible to quantify the probability of success for the project as related to political factors.

References: Gupta and Wilemon, 1990; Allen, 1970.

9.5.2.12. Management and Control

The project plan will typically include a description of the management and control of the project once it is underway. Covered are topics such as measurement metrics, type of variance reporting and responses, audit plan and a description of major and submilestones and decision points.

9.5.2.12.1. Measurement Metrics. Benchmarking participants voice agreement that critical success factors should be identified in the planning stage of the project. The factors originate in the original discussions with the project customer about their needs and the project deliverables. The team can also review other similar proj-

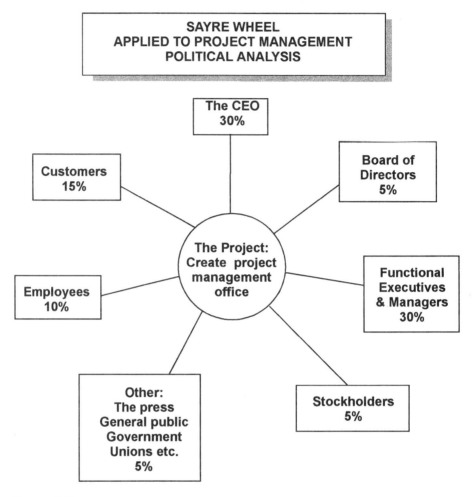

FIGURE 9.5

ects or benchmark with other groups to determine measurement methodology.

> **Standard:** The best practices project methodology identifies customer requirements and measures project performance against those requirements.

Once the project is underway, measurement metrics should be in place to determine progress against the critical success factors. Benchmarkers state that sometimes the measurement of project performance is subjective, nebulous and difficult to define, but it is still necessary. Where possible, such as when measuring cost and time, and conformance to functional and technical specifications and resource usage, the variances can be more precise.

Benchmarkers also emphasize that measurement methodology adds time and cost to the project. Often there is a tendency to focus too much on measurement metrics and tools and to forget to lead and manage the project. Consequently the amount of detail should be appropriate to the original statement of customer needs and requirements. Part II of this book contains a detailed discussion of various performance measurement metrics used by benchmarking participants.

9.5.2.12.2. Variance Reports. The manner in which the cost budget and schedule are monitored is through the use of variance reports. The project plan will describe in appropriate detail how variance reporting is to be conducted. Most variance reports are detailed in nature. For example, when looking at the cost variances the report will show the original budget and actual amounts spent. Negative or unfavorable numbers are shown in brackets. Particularly large deviations from the plan may be highlighted to make the scanning of multiple pages of variances more easy.

9.5.2.12.3. Audits. The arm's length, independent, and objective evaluation of the project is the audit. The project plan will describe when audits are to be performed and areas to be investigated. Generally the audit will seek to determine if the project is on cost and schedule and if the technical and functional specifications are being attained.

9.5.2.12.4. Milestones/Decision Points. The 60 companies participating in the benchmarking forum place high reliance upon the use of milestones and decision points as triggers to evaluate project performance. The project plan will identify the project wide milestones as well as milestones for subprojects. Where go/no go decisions are required, they will be detailed as well.

9.5.2.12.5. Scope Change Management. The procedures for managing changes to the project scope are detailed in the project plan. Often scope changes represent additional revenue and the primary source of profit on external projects. They also tend to make project performance difficult to measure, change the nature of the project, and result in cost and schedule overruns. Invariably, scope changes will be formal in nature and require formal approval in the form of signatures.

9.5.2.13. Signatures

When the project plan is complete, signatures are required to authorize the execution phase of the project. Generally the signatures of the customer or project sponsor are included. Often, particularly for internal projects, the signatures of other stakeholder groups are included as well.

9.5.3. Execution and Control Phase

During the project execution and control phase, the superior project organization ensures that the project team implements the project plan and engages in the day-to-day management of the project. Emphasis shifts from plan development to getting the work done through people focused leadership and management. The project plan and its associated components become the tools used to monitor, measure, and control project performance as reflected by the schedule, budget, and functional and technical specifications. Included are monitoring of the project scope, managing project changes and responding to risk events as well as opportunities. Progress toward achieving goals and meeting deliverable objectives receives constant attention.

All prior efforts expended on project initiation, selection, and planning are for nothing unless the project is executed speedily, efficiently and effectively. Schedule targets and milestones are met only if the project is executed speedily. Budget and cost targets fall by the wayside unless project execution is efficient. Effectiveness is attained when all the elements of the execution phase result in the presentation of project deliverables to the satisfaction of the customer and stakeholders. In the final analysis, accolades (or blame) are awarded to the project manager who successfully executes the project, not to the project plan, the charter or tools and templates used.

9.5.3.1. Kickoff Meeting

Participants in the benchmarking forum believe the kickoff meeting has high positive correlation with project goal achievement. At least two major international companies consider it so important that they bring globally dispersed project team members to a central location for the kickoff meeting. They avow that it is the one time in the life of the project when all the team members should be physically together.

Many kickoff meetings include project stakeholders as well as team members.

The importance of the kickoff meeting is supported by the research. Those who study the impact of formal ceremonies say that they play an important part in redirecting people's lives. They shift team members' thinking to a new mindset and mental track. An example is the impact of the wedding ceremony. People who have engaged in long term courtships generally agree that the nature of their relationship changes as a result of the wedding ceremony.

The same is true of the kickoff meeting. The kickoff meeting emphasizes team building by gaining commitment to project goals and deliverables, reviewing the project charter, discussing individual assignments, and addressing any of the other myriad elements of the project considered important.

Standard: **The best practices project methodology includes a kickoff meeting.**

Participants also observe that the attendance at the kickoff meeting is a gauge to determine stakeholders' and the organization's view of the project. If it is poorly attended, it could be an indicator that the project has been incorrectly promoted or that people fail to see its importance. One group rescheduled the kick off meeting to start the new project management office four times. They were faced with the conclusion that there was a need for more promotion of the value of project management to other functional areas of the company before the project could expect a high probability of success.

9.5.3.2. Importance of Project Control

The *quality* of project control is a factor that distinguishes the superior project manager from others. It also represents potential for sizeable savings for organizations that use professional project managers. For example, Ingram's investigation of 60 failed projects determined that 70% came in materially late, over budget, or failed to meet the client's expectations. The direct and associated costs of failure *averaged eight to ten times the original project budget.* Research of projects investigated by the Standish Group concludes that 86% of software projects come in late and/or over cost. Further, almost all projects in the study incurred significant scope changes. Success rates varied by company size. In large companies, only 9% of projects came in on time and budget. Smaller companies recorded better results. Seventy-eight percent of their software projects were deployed on time and budget while maintaining approximately 75% of their original features and functions.

> **Standard:** **The best practices project methodology includes the use of variance analysis as a tool to control deviations from the project plan.**

9.5.3.3.   Priority of Time, Cost, and Specifications

Benchmarking participants were surveyed to determine the relative amount of emphasis placed by customers on attaining time, cost, and specification goals. Respondents agreed that meeting time targets was consistently *most important* although they were closely followed in perceived importance by meeting cost targets. Functional, technical, and quality targets were a distant third.

To demonstrate the reasoning of respondents, consider the construction of a new home with a forecasted completion date of December 15th. Assume the new homeowners are excited to be planning the Christmas holidays in their new home. They even invite their families to join them. On December 1st imagine that the contractor informs the homeowners that the house will be one week late. After the initial distress, a typical reaction of the homeowners would be to investigate options available to let them move into the home on time. Likely the contractor would need extra funds to pay workers overtime and perhaps hire a couple of additional laborers. In many cases, the homeowner would be willing to pay extra (i.e., compromise the cost target) to satisfy their time schedule desires.

Now consider the perceived importance of meeting technical and functional specification targets. In the new home situation, assume that the contractor has committed to meeting the delivery targets as a result of receiving the additional funding. But now the contractor introduces the additional problems that the ornamental trim in the third bedroom will not be finished nor will the back fence or the landscaping in the yard. In response, most homebuyers would agree to compromise or delay these items in order to move in on time.

9.5.3.4.   Time Control and Monitoring

The world of the superior project manager is dominated by time important are starting meetings and keeping appointments precisely as scheduled, completing work packages when due, and remembering that achieving time targets is generally considered the most important measure of performance by project customers.

Time management is clearly an area where the best practices of the superior project manager have potential to generate positive results. In support of this assertion, the Standish Group studied the success of projects meeting time targets. They determined that only 14% of projects came in on time or recorded less than a 10% time overrun. Approximately 46% experienced time overruns of 100 to 400%. The average overrun was 222% of the original time estimate.

9.5.3.4.1. Time Variance Measurement. The superior project manager monitors and controls the project schedule through observation, use of milestones, percentage complete, and earned value calculations.

- **Observation.** Benchmarking participants emphasize that the best way to measure project progress is to physically observe and verify. All other methods involve unsubstantiated reports and delayed communications. Observation generally results in immediate feedback of project status. Deviations from the plan are quickly identified and action taken to correct the variance. Without observation, benchmarkers report that there is a tendency for team members to overestimate the degree of completion of individual work packages. Hence the phrase, "We are 80% complete with 80% left to perform."
- **Milestones.** One of the most important control methods in project management is to insert frequent milestones in the schedule. Every for-profit organization in the benchmarking forum uses milestones as the *primary* method for monitoring and controlling schedules and project time performance.

  Milestones are convenient points that encourage the team to evaluate the project and measure its performance before proceeding to the next step and incurring additional expenses. Upon arriving at the milestone, it is a simple matter to observe whether the project is ahead of or behind schedule. For example, assume that a new home is scheduled for completion on December 1st. If December 5th has been reached and the house is not complete, it is intuitively logical that it is five days late.
- **Percentage complete and earned value.** Sometimes milestones are too infrequent or inappropriate to use for schedule evaluation. On projects with overlapping project phases there may be few convenient places where milestones can be placed.

The same is true when taking a multifunctional approach or when there are numerous subprojects and the various elements progress at different rates. Sometimes an evaluation of project status is required *between* milestones. In all these situations, percentage complete or "earned value" progress calculations represent more understandable measurement devices.

Benchmarking participants report the use of two types of percentage complete calculations. Both are simple to calculate and easy to understand. One compares the number of *work packages completed* with the total amount planned for the project. The other compares *funds expended* with those scheduled to be spent for the entire project. For example, assume the project is the construction of a new product prototype. Scheduled time to the prototype completion milestone is six months. The number of work packages in the total project is 600 and budgeted funds are $1,200,000. If a status report is requested at the end of month three (half way through the project), it is a simple matter to count work packages complete and add up the money spent. In this example, if 300 work packages are complete (50% of the total) and $600,000 has been spent (50% of funds budgeted), the project is on schedule.

A somewhat more complicated form of percentage complete calculation is earned value. It compares the budgeted cost of work performed to date (BCWP) with the budgeted cost of work scheduled to date (BCWS). The formula is BCWP – BCWS to obtain a dollar amount, or BCWP/BCWS to provide an index or percentage. It is different conceptually from the more simple percentage complete formula because it works only with budgeted amounts rather than actual amounts.

Earned value is primarily used by the U.S. government and associated contractors. A survey of 60 participants in the Top 500 Benchmarking Forum concludes that *none* use earned value for schedule management. The major complaint is that earned value is complicated and difficult to understand.

Note that earned value consists of two basic elements. The first is *schedule variance* and the second determines *cost* performance. The costing portion compares budgeted with actual work package costs. The costing portion reflects the conventional cost accounting approach and is widely used and accepted within for-profit organizations. It is discussed in the Section 9.5.3.5.

9.5.3.4.2. Team Emphasis on Time. Motivational techniques are applied to maintain team focus on the schedule. Included are peer pressure, status reviews, and the posting of team and individual performance.

References: Ingram, 1998.

9.5.3.5. Project Financial Control

The second area where the superior project management has major impact on project performance is control of the budget. It is the financial bottom line for the project and is comparable to executive efforts made to attain a profit for an entire company. Research indicates that project financial control is an area with major potential to benefit from project management best practices. The average cost overrun for projects studied by the Standish Group was approximately 190% of the original cost estimate.

Conventional costing concepts from the world of accounting are directly applicable to project management. The procedure is as simple as comparing the amount budgeted with the amount actually spent. The result is the variance. There are two types of variances commonly used: the *amount* of any resource used (for example, the number of people) and the *unit cost* of each resource (for example, the estimated cost per hour of a person).

The difference between each of the costing systems is the type measurement performed. A summary of each follows.

9.5.3.5.1. Earned Value. The costing portion of earned value compares *estimated* total work package costs with total *actual* work package costs. No distinction is made between number of resources used or unit costs for each resource. The formula to determine earned value costing variance is budgeted cost of work performed *minus* actual cost of work performed (CV = BCWP − ACWP) to find a dollar amount, and budgeted cost of work performed *divided* by actual cost of work performed (CPI = BCWP/ACWP) to determine a percentage or index of cost variance.

The simplified nature of the earned value cost calculation lends itself to estimating completed costs and efficiencies required to meet targets. The estimate of the cost of the completed project, assuming that current cost overruns will continue, is determined by dividing the cost index by the original total budgeted amount for the project (EAC = CPI/BAC). The degree of efficiency needed to bring the project in on the original cost target is determined with the formula TCPI = (BAC − BCWP)/(BAC-ACWP).

9.5.3.5.2. Activity Based Costing. Most large projects lend themselves to activity based costing. The cost of an entire activity is measured over the period that the activity is in progress. It is particularly appropriate if a project spans multiple accounting periods. Traditional accounting is geared toward closing the books at the end of major periods such as year end. With activity based costing, the costs and revenues continue to be recorded over the life of the project.

9.5.3.5.3. Life Cycle Costing. The entire life of a *product* is the measurement standard for life cycle costing. It is often encountered in the project management environment, particularly in association with new product evaluations. Life cycle costing differs from other costing methods because it generally includes revenue as well as material and labor costs. When evaluating the life cycle of capital goods, machinery, and many consumer products, there are sizeable cash flows after the sale of the product. Included are such items as repair and service components, maintenance, potential to move up to larger models, added features, trade-in potential, and salvage value. Numerous products have greater cash flows *after* the product has been sold than before.

Life cycle costing can have dramatic impact on new product selection and strategy. For example, some computer printers are sold at low prices with little if any profit made from the product sale. The marketing strategy is to generate lifetime profit from the sale of toner.

9.5.3.5.4. Job Order Costing. Job order costing applies costing concepts to jobs that have a specific beginning and ending and include clearly definable units of production. In a project environment it could pertain to the activities or segments between each milestone or individual work package or even the total project.

Job order costing measures the cost to produce a group of products or services. For example, a "batch" in a sporting goods factory might be 50 baseball bats. Total material and labor costs to build the batch of bats would be recorded. To determine the cost of one bat would entail dividing the total cost recorded by 50.

9.5.3.5.5. Process Costing. The difference between job order and process costing is that job order costing measures the cost of a group of *units of production* and process costing measures units produced in *a specific period of time*. Process costing is used in operations such as mining and petroleum production. Some observers judge that process costing has little application to projects. However, if a project has few milestones, over a period of time it begins to take on the attributes of a process flow operation. Further, if one considers earned value calculations, it will be noted that the calculations apply more to projects with continuous flows rather than those with many small, segmented milestones. In a way, the calculations associated with earned value could be considered an attempt to measure progress in a process flow environment.

9.5.3.5.6. Standard Costs and Variance Analysis. During the project planning process, the costs for each of the tasks comprising project work packages are estimated. These estimated costs become the "standard" costs of the project. They have counterparts in nearly every industry. In the auto service business they are termed flat rates. The primary reason they are used is because it is easier to use an estimate than to calculate actual rates, and actual costs are not known. Once the project is underway, actual costs begin to be accumulated. Comparing the actual costs with the estimated or standard costs results in variances.

9.5.3.6. Scope Management and Control

Alterations to the project that affect the schedule, budget, functional and technical specifications, or any other aspect that involves the expenditure of funds or resources are labeled "scope changes." Benchmarking forum participants report that the project manager is under constant pressure to change the project. Sources of the change requests are numerous: they can be oral or written from project managers themselves, team members, or external stakeholders; they can be legally mandated; or they could be corrections for errors and oversights. The process is always approached in a formalized manner. Even so, there is a tendency to make informal changes even when formal procedures are in place.

Scope changes are a double-edged sword for the superior project manager. On the one hand, scope changes negatively impact project speed, efficiency, and effectiveness. They generate alterations to the budgeted cost of work scheduled, the budget at completion, the day-to-day schedule including periodic milestones, the technical and functional specifications, and even deliverables and project goals.

In all cases, scope change makes it more difficult to measure performance, monitor progress related to the original project plan, and to compare current progress with prior efforts and milestones. Projects become more nebulous, ever-changing, and harder to define.

A problem stated by benchmarking participants is that often the project manager *allows and even encourages* scope creep and project changes. Usually the project manager is making an effort to be cooperative and to keep the customer happy. In other cases, the scope changes may be self-serving in nature. Ingram's research identified numerous examples where project managers used uncontrolled scope changes to reduce accountability or to continue projects indefinitely when they should have been terminated.

On the other hand, scope changes represent opportunities to generate incremental revenue and profit, to add value, to adjust the project to changing conditions, to incorporate improvements, and to correct oversights. Scope change is a key component of the sales strategy in many organizations. The companies bid projects at low prices to obtain the contract with the intention of making the profit from the predictable scope changes.

One thing is certain, scope changes will occur. Zangwill's research found constant pressure from stakeholders to change the scope of the product. He determined that the most successful project teams applied discipline and formal procedures to control such cost increasing modifications. Project managers were trained to manage change in a formalized fashion and to enforce the formal scope change procedure.

The process of ensuring that the project is meeting the scope objectives and conforming to the project plan requires constant monitoring by the project manager. For example, to prevent the tendency of team members to make unauthorized changes, the project manager might ask at each periodic status meeting, "Have any changes been made?" This procedure serves to bring the impact and consideration of the effects of changes down to the individual project team members.

> **Standard:** **The best practices project methodology manages project scope changes as a formal process.**

9.5.3.6.1. Scope Change Process. The scope change control process is defined during the planning stage and communicated to project customers and other stakeholders. In contractual project relationships, the scope change process should be spelled out in the contract. Described is the paperwork that initiates the change and description of the approval process. The scope change process always involves obtaining customer signatures.

The management of scope change involves analyzing the change to determine its impact on the project plan. Often the process incorporates other stakeholders and team member evaluations. Many organizations have change control committees composed of members from the various functional groups affected by the project. The committees approve changes as well as ensure that all stakeholders understand the effects of the change. On large projects authorization may be required by as many as five change control boards. One board might review market impact, another manufacturing implications, another serviceability, and another price and cost effects.

9.5.3.6.2. Encourage Changes Early. Benchmarking participants stress the importance of emphasizing to customers that changes made early in the project life cycle are generally easier than making the same changes after the details become set in concrete and project execution commences. It will become progressively more difficult to make scope changes as the project progresses.

Most elements of the product or service are locked in early in the process. The basic functional and performance characteristics, resources to be used, and expected outcomes are typical foundation decisions. Once these fundamental decisions have been made, any change affects the entire project. At later stages, all the elements of the project are so interrelated that even a small change can affect several aspects of the project. At some point, there is almost nothing that would be described as a small change.

Zangwill concluded from his research that delaying changes through each subsequent phase multiplies the cost of the change by a factor of 10. His calculation was as follows:

During design	$1,000
During design testing	$10,000
During process testing	$100,000
During test productions	$1,000,000
During final production	$10,000,000

Zangwill justified his logic by quoting projects where changes in testing cost 13 times more than similar changes made in early design. He found that changes after customer installation increased cost 92 times. In the projects researched, the concept development stage accounted for *1% of total project cost but determined 70–80% of total life cycle cost.*

References: Ingram, 1998; Zangwill, 1992.

9.5.3.7. Project Value Analysis

Value analysis refers to the formal process of eliminating cost from the project or the project's product and service. The objective of value analysis is to accomplish the goals of the project, product, or service at the minimal cost.

This process is necessary because often cost is not the major consideration during the project planning process. Generally planners focus on including every element, work package, and activity. The emphasis is on establishing time and budget definitions. If a new product is being designed, the perspective is usually on performance and meeting specifications. During all these activities, cost is a secondary element. If the project is of long duration, say several years, the constant influence of external economic factors encourages higher labor rates, more expensive materials, and increases in the overall cost of doing business. There is a tendency for costs to increase even more as unforeseen events occur and schedules change. Often in the haste to get underway or to design a product against a deadline, quality features and activities may be added that do not contribute to the product or overall project suitability.

From this view, the extra quality, detail, and features represent wasted expenditures. More over, the cause of the cost may result in continuation of the problem. The value analysis is a formal process to review the project and/or its associated product or service with the objective of reducing costs without impairing efficiency, effectiveness, or suitability of the product or service.

9.5.3.8. Asset Stewardship

Project managers are entrusted with sizeable amounts of assets and funds. The concept of stewardship recognizes the fiduciary responsibility of the project manager to protect those assets. Most commonly, in the world of finance and accounting asset stewardship includes protecting the project's assets against theft. Applied are the numerous accounting controls that serve notice that the assets are being monitored and protected.

9.5.3.9. Communications

As discussed in prior chapters, the communications process is an important component of project success. All benchmarking participants use a combination of formal and informal communications procedures.

For reporting project status to stakeholders, best practices project managers use the simplest template that meets the information needs of stakeholders. Templates summarize important project information on a single page. Some benchmarking participants simplify the template even more. The streetlight format presents critical reporting activities (e.g., schedule, cost, specifications) in a "red, yellow, green" visual format. Other companies use a visual "project control panel" where the important tracking activities are presented as gauges.

All participants encourage periodic meetings to discuss project status. Frequency varies according to the nature of the project. On critical and politically sensitive projects, some report as many as three status meetings every day. Others have as few as one meeting per month.

9.5.3.10. Formal Reviews and Audits

An audit provides project teams as well as stakeholders an independent and objective evaluation and information about the project. Its objective is to investigate the project's performance in meeting schedule, cost, and specifications as well as being on track in achieving objectives and deliverable targets. Captured are both the good and the bad aspects of project performance. The audit or review provides a safety valve for identifying and communicating project problems to stakeholders. Typically, the audit is performed by the project organization or other group external to the project itself. The process is described in detail in Part II.

9.5.4. Project Closing and Termination Phase

Project termination includes the critical activities of presenting the final deliverables to the customer, preparing and executing the project termination checklist, placing historical information in a knowledge library (i.e., lessons learned, project journals, project charter, and plans), communicating the benefits of the project to stakeholders and the host organization, and conducting the termination celebration. In some cases, project follow-up activities include warranty and service provision and monitoring of project results.

Many seasoned project executives judge the termination process one of the most critical elements of successful project leadership. The manner in which it is performed directly impacts the residual attitudes of stakeholders about the project and the project team. It is also the phase where numerous pressures encourage the project manager to be hasty in its execution. Benchmarking participants report that it is common to review project plans that fail to include termination costs or the steps necessary to terminate the project.

When long-term projects are concluding, and particularly when team members are unsure of their futures, stress is high. Researchers suggest that the impact on participants' emotions is on a level with that associated with other life crises such as divorce, suffering the death of a loved one, or getting fired from one's job. For example, the project manager for an Olympic Committee told an Academy of Management group that approximately 5,000 volunteers worked up to five years preparing for the Olympics. After their years of effort, the Olympics reached its conclusion in a crescendo of activities. According to the project manager, the reported emotional problems and even suicide rate among participants was dramatically higher in the "letdown" months following termination.

Benchmarking participants report that emotions are manifested in other ways. For projects that are less than successful, finger pointing and accusations can be expected. For those that are wildly successful, jealousy and resistance from entrenched functional areas of the organization often temper the elation of the project team members. In every case, the astute project manager can expect nontypical reactions and responses.

Another problem reported by the benchmarking participants is that as termination nears, project team members including the project manager become subject to a "termination" state of mind. It is similar to the emotions that arise during a five day course or training meeting. On the last day, most people are preoccupied with thoughts about what will happen as soon as the meeting is over. Almost everyone is wishing there could be a way to accelerate the termination process and end the meeting early. Someone will usually even encourage the leader to do so. In these cases the superior project manager and team members will force themselves to be professionals and to complete all the scheduled activities.

9.5.4.1. Presenting the Deliverables to the Customer

All the previous phases of the project culminate at one point: the presentation of the deliverables to the customer. Benchmarking participants agree that the final presentation of deliverables should take the form of a formal ceremony. Even for projects where deliverables are presented to the customer over the entire project life, there should be a ceremony for the final delivery. Often the final delivery is symbolic since the actual product or service might already be in use by the customer. Examples are the final walk-through for new buildings and the cutting of a ribbon for highways. The formality of the final delivery signifies to the customer and team members that the project is over, its responsibility is being conveyed to the customer and a new direction in the relationship is commencing.

Standard: **The best practices project methodology emphasizes formal presentation of the final deliverable to the customer.**

9.5.4.2. Preparing and Executing the Project Termination Checklist

The termination of many projects is a sizeable undertaking. To aid in the process, most project managers list activities to be accomplished in a checklist format. The checklist ensures that all activities are performed in a professional manner and no shortcuts are taken.

Standard: **The best practices project methodology includes a termination of checklist to ensure all termination activities are performed.**

The checklist is a reflection of the original project plan. It ensures that all activities are completed and closed. The sidebar shows a sample checklist for the accounting and finance elements of a typical project. Resources and assets are inventoried and returned to their proper destinations. Accounts are closed, all monies are accounted for, and property and equipment are inventoried. Many project organizations require an audit of the accounting and financial records.

Other checklist topics include recognition and closure of scope changes and the review of functional and technical specifications. All commitments made to the customer are reviewed to ensure that they have been fulfilled. Contracts and agreements with vendors are terminated. Purchasing and other functional areas are notified of the project's termination.

Project Termination
Financial and Accounting Check List

1. Financial and accounting documents have been closed and recorded.
2. Final charges and costs have been audited.
3. Audit has been completed of all the financial and accounting records.
4. Final project financial report has been completed.
5. Receivables have been collected.
6. Payables have been paid.
7. Vendor contracts have been reviewed to ensure all terms have been satisfied.
8. All contractual deliverables and completion dates have been met.
9. Work orders and contracts closed out.
10. Financial reporting procedures terminated.
11. Financial reports submitted to pertinent stakeholders.
12. Pertinent licenses and permits terminated.
13. Nonessential project records destroyed.
14. Essential project records stored.
15. Electricity turned off, doors closed and locked, goodbyes said.

9.5.4.3. Placing Project Documents in a Knowledge Library

The superior project organization contributes to the organization's body of knowledge. Information that could benefit and improve the performance of future projects is filed in an appropriate library. Future project managers can refer to the library of project charters and detailed plans, tools used, lessons learned, and project journals to serve as templates and assist in developing new projects. Some benchmarking participants have such extensive libraries that new project managers can pick similar completed projects and model the new project after the old. Whether one is reviewing the work breakdown structure, the project schedule, prior budgets, specifications, journals or lessons learned, using an existing project as a template increases the new project's speed, efficiency, and effectiveness. Future schedules and costs become more accurate, risks and pitfalls are more readily identified, and all project phases become more efficient because the project manager has access to and builds upon prior experiences.

> **Standard:**　**The best practices project organization contributes to the organization's project knowledge base by maintaining appropriate project documentation (lessons learned and journals, the charter, and plans) in a library or depository.**

The superior project methodology includes ways to evaluate the project's successes and failures. This is done through the process of encouraging project managers to maintain a project journal throughout the life of the project or compiling lessons learned at the end of the project. From these, future project managers gain access to knowledge acquired during the transmission of each successive project.

Benchmarkers advise that the critical self-analysis associated with lessons learned and project journals is a sensitive issue in many organizational cultures. Any tendency toward finger pointing or critical evaluation of the performance of specific groups can be unpopular. In some organizations, public acknowledgment of one's failings in leading the project can have negative career consequences. Consequently, the project organization assists and guides project managers in carefully selecting appropriate wording before distributing or preserving lessons learned and project journals.

9.5.4.3.1. Lessons Learned.　At the end of the project, the project team develops a list of lessons learned. A few benchmarking organizations wait four or five months after completion to ensure the availability of information. The logic is that the project team can be more independent, objective, and more accurately identify things that went wrong or were executed in a superior fashion. Benchmarkers recommend that lessons learned be reviewed at the kickoff meeting or during the planning stages of new projects.

Some project groups use a generalized lessons learned template to guide a structured review of the project. Schedule, cost, and specification changes are examined. Risks encountered are recorded. Major variances from the project plan are discussed and noted. Personnel, political, stakeholder, and environmental issues are reviewed as are incorrect assumptions made during the planning stage.

Benchmarkers state that the tendency of lessons learned is to focus on problems. However they stress the need to emphasize the positive aspects of the project since the greatest returns result when people understand successes and why they occur. Best practices project organizations recommend that, where possible, lessons learned should be quantified. Identification of the cost of errors and problems provides more precise weighting for future risk items. Variances are sometimes ranked according to the amount of financial impact on the project.

9.5.4.3.2. Project Journal. Some best practices project managers maintain a journal or diary during the project's life. A typical journal would contain a short summary of knowledge acquired on a daily basis and is comparable to the captain's log on a ship. It is written as though the project manager were communicating with a close friend or relative who is to be the next project manager. On some project teams, each team member is encouraged to keep a journal.

An abbreviated form of a journal results when the project team develops interim lessons learned for each project assessment review, milestone, or audit. These are then compiled into a final project journal or lessons learned document.

Users of journals profess that they offer advantages when compared to lessons learned. They often are more compatible with the organizational culture. They tend to relate events and show the chronology of situations. They are less inclined to make sweeping conclusions and generalizations.

9.5.4.4. Communicating the Benefits of the Project to Stakeholders and the Host Organization. Measurement and communication of the project group's performance is crucial. To the project organization, project manager, and team members, the benefits and positive results of the project effort seem clear; but the portrait is often opaque to senior executives, stakeholders, and others in the organization. Benchmarking participants unanimously agree that for project management to continue to be recognized as a profession of value to organizations, there must be a clear linkage between its application and improved organizational project goal achievement. To accomplish this, it is necessary to measure and communicate project results and performance. Part II of this book details the strategic importance of the process and the ways that benchmarkers have communicated project benefits.

> **Standard:** **The best practices project methodology includes measurement and communication of project contributions to stakeholders and the organization.**

Listed in the following are various approaches benchmarking organizations take to measure project benefits that were not covered in Part II. Any or all of these measurement methods can be tailored to articulate project benefits with the host organization's goals.

9.5.4.4.1. Value of Professional Project Management. The use of professional project managers should improve project performance well in excess of its cost. This issue is of particular importance because the professional project manager is generally more expensive than the nonprofessional. Consequently, best practices project groups measure the cost of professional project management. The cost is then compared with project output. One measurement tool is to simulate what would have happened if the project had not been managed in a professional manner. Several project organizations have a line item "cost for project management" or "construction management." They measure the percentage amount of the project that is management. Project management groups quote that the cost of professional project management ranges from 1.5 to 7% of total project cost. On some government-managed projects, it is common to observe costs of project management as high as 40% of total project cost.

9.5.4.4.2. Impact on Profits. Most organizations are bottom line oriented. The question is almost always asked, "What did the project contribute to the organization's profit, goals, and strategies?" Specific metrics might be increased sales and revenues, reductions in cost, improvement in organizational efficiency, and return on investment.

9.5.4.4.3. Accuracy in Achieving Time and Budget Goals. Several best practice organizations evaluate projects on the basis of accuracy in meeting goals. For example, a project might be under or over budget by 5%. Either result is equally acceptable and would receive the same management response.

9.5.4.4.4. Efficient Use of Resources. Projects consume resources. One objective of the superior project manager is to minimize wasted labor, materials, and other resources and to maximize output. On some projects it is possible to measure productivity and efficiency by comparing work output and resources used. Project work output can be compared with cost and/or resources used such as labor. One danger of this method is that it encourages cutting costs to the point that output is reduced as well.

9.5.4.4.5. Success in Meeting Commitments. The question can be asked, "Did we meet the commitments we made?" The project team can review the promises made during the planning stage and throughout the project. These can be evaluated in relation to scope changes to determine the degree of success in keeping the commitments.

9.5.4.4.6. Customer Satisfaction. There is general agreement that one dominant measure of project performance is customer satisfaction. The superior project manager identifies customer requirements and measures project performance against those requirements as determined by customer satisfaction. Measurement tools usually consist of informal meetings or formal surveys. Note that benchmarking participants stress that customer satisfaction should be measured continually as well as at project termination.

9.5.4.4.7. Client Retention. The proof of customer satisfaction is retention as evidenced by repeat business and a strengthening of the relationship.

9.5.4.4.8. Stakeholder Satisfaction. When taking a stakeholder view of team performance, team success becomes a multidimensional issue. Issues in addition to the project's success need to be taken into account. Different constituencies have varying degrees and definitions of superior performance. Interest groups or individuals that can impact the perception of project success can be evaluated. Where there is evidence of satisfaction, this should be communicated.

9.5.4.4.9. Reduction in Time to Market or Completion Time. One measurement of project performance is to compare the former time to complete projects or "lead time to market" with those achieved by the current project. Generally, one of the most dramatic results of professional project management is the reduction in time to complete projects.

When a new product is involved, the practical benefit of reducing project time is that the product enters the market sooner. Generally, the historical time to develop new products is well known within an organization. It is probably never precisely quantified, but experienced managers and executives will have a general idea of traditional lead time to market. For example, one company in the Top 500 Project Management Benchmarking Forum judged that before the project group was started, new product development time to market was 52 months. The projects being managed by professional project managers reduced lead-time to market to an average of 18 months.

Benchmarkers report that the value of reducing time to market by one day is phenomenal. Each day time to market is reduced adds one more day of sales. In the preceding example, by shortening lead time from 52 to 18 months, sales occurred nearly three years earlier than they would have without the project approach.

Benefits accrue to the company in other ways. By generating sales sooner, money is made available for other investments. Furthermore, shortened time to market means that market position is established earlier. Incremental sales are obtained and higher profits occur by applying pricing techniques to capitalize on the market niche, limited competition, and initial high demand product for the new product. Finally, marginal products are identified earlier in the design process.

If one considers solely the impact of the time value of money, the results are dramatic. The money generated by the product sales can be invested in other income producing opportunities.

Reducing lead time to market offers other, more subtle benefits. Its effects are similar to increasing inventory turnover. A high inventory turnover means that the investment in inventory is small compared to the amount of resultant sales. By reducing time to market, the same number of people can complete more projects. Where previously a person might be involved in a project for, say a year or so, now they will complete one in a few months and then move on to another. The cash investment in each project is reduced. Savings occur from (a) making project managers and team members available for other, higher potential projects and (b) eliminating the investment associated with the project development expenses for the extra months of development time.

Reducing lead-time to market gives the company capability to react faster to market changes. The organization has the confidence of knowing they can respond to market changes faster than competition and consistently be first in the market. The first products to be introduced enjoy the benefits of high demand and low supply or scarcity of the product. Consequently, an astute pricing policy will capitalize on this set of circumstances and maximize profits. As the market matures and competition begins to enter, then pricing becomes more competitive and profit margins decline.

Products that are early in the market also have potential to carve out a niche and establish a stronger competitive position. Latecomers are relegated to the reduced and declining profits reflective of the maturity and obsolescence stages of the product life cycle.

9.5.4.4.10. Impact of Problem Identification and Corrective Action. One advantage of continuous measurement and monitoring of project performance is that marginal projects are identified and culled earlier in the development process. One project management group indicates that they identify and terminate projects on an average of three months after execution commences. At the inception of their performance measurement efforts, it took approximately one year to terminate a clearly losing project. Other groups report that they measure the time to identify problems and the number of project problems corrected.

9.5.4.4.11. Value of Quality. Benchmark participants report that measuring the cost of quality is a noble goal, but difficult to do. A few best practices companies have anecdotal estimates of its value. One company formally queries customers about the degree of satisfaction with project quality. Benchmarkers agree that even with the measurement problems the value of quality should remain a subject that receives discussion and is a part of the measurement component.

9.5.4.4.12. Overall Portfolio Benefits. Some project organizations attempt to show the value of the individual project as it relates to the entire portfolio of projects. Most use a baseline of project return versus risk delineated at the point professional portfolio management commenced. They then compare the baseline with the current portfolio of projects.

9.5.4.4.13. Aggregate Benefits of Projects. All of the measurement methods used by a project organization can be aggregated over a period of time and numerous projects. The totaling of measurements gives the project organization capability to measure improvement and to set goals for future projects. Some participating organizations roll up the aggregates of all their projects for the year. They show the results of each specific project as well as all projects in total. Status and performance measurements are graphically portrayed for all stakeholders to evaluate.

References: Ibbs, 1997; Ingram, 1999; Ancona and Caldwell, 1990.

9.5.4.5. The Termination Celebration

Usually at the conclusion of the project is a termination celebration. It is often combined with the presentation of deliverables. For large projects it may include only team members and stakeholders. The termination celebration serves a similar purpose to that of the kickoff meeting. It mentally redirects people from a project execution mode to termination and moving on to a new endeavor. The termination celebration usually includes formal or informal recognition for services performed, a review of events where bonding occurred, overview of lessons learned, and humorous incidents. In cases where the project has been a failure, the termination celebration may be skipped.

9.5.4.6. Project Follow-Up

After the official termination of the project, occasionally activities remain to be performed. Product market performance, service and durability experience, and warranty costs may be tracked and compared with estimates.

9.5.4.6.1. Warranties. Many projects produce a product or service. Often companies agree to provide free service on units that are defective or fail to perform properly. When these costs, or warranties, are expected to be minor in nature the project team usually turns over their responsibility to a functional area of the organization. However, if the warranty costs are expected to be sizeable they may have an impact on the perception of stakeholders and customers about the overall success of the project. In these cases, their responsibility may remain for a short time with the technical staff (i.e., the project team) that was responsible for the product's design.

Warranty costs are usually estimated during the project development stage and are based on past experience. In situations involving a totally new product or service, the warranty estimate can be based on an analysis of individual components and testing experience. The preplanning for warranty cost gives the company time to develop a mitigation strategy for expected and unexpected problems. Customers can be sold a service contract, which results in shifting some of the financial burden. In effect, the service contract is an insurance contract executed with the customer. Education programs can be developed to prepare customers for the expected results and maintenance procedures. Finally, warranty preplanning provides information needed to budget the cost into future financial statements.

10

Knowledge Management: Providing Tools, Standards, and Training

The previous chapter discussed the knowledge management function and importance of developing and implementing a structured and predictable methodology and templates as well as transferring knowledge through the knowledge library. This chapter details the key role of the project organization in increasing overall project speed, efficiency, and effectiveness through provision of a superior tool kit, by setting high standards of performance, and by training project managers in project specific skills.

10.1. PROVIDE STATE-OF-THE-ART PROJECT MANAGEMENT TOOLS

Project management tools constantly change and improve. A role of the project organization is to remain aware of, evaluate, test, recommend, and make available the latest technology and software. Currently being emphasized is the internet, intranet, email, and various specialized software packages. All agree that the technology is changing almost by the day and requires constant attention by the project organization.

Software and Tools Selection Criteria

Benchmarkers have developed checklists of minimal standards
or criteria for software and tools selection and evaluation. One
such list is as follows:

- Scalable.
- Globally compatible with key languages, needs of glob-
  ally dispersed project teams.
- Web enabled. Can publish, interact, & change.
- Desktop. Must be compatible with browser.
- Must be customizable.
- Linked to planning, scheduling, resource control, and
  time.
- Searchable or archivable.
- Detailed security function.
- Dial-in capability.
- Works with various programs.
- Ease of use and easy to learn.

As evidence of the importance placed by benchmarking participants on remaining current, a portion of each benchmarking forum
is allocated to having participants discuss and describe new tools and
their experience in using the tools. In addition, almost all of the
benchmarking organizations train project managers in the use and
navigation of web pages. Training is particularly important in the
global environment where languages vary.

Benchmarkers generally recommend specific preferred tools for
organization wide use. The side bar shows a checklist used by one of
the benchmarking organizations for evaluating project related software and tools. In all cases, they recommend that tools and software
standards be flexible for application in various countries and meet
the needs of individual projects and portfolios. Consequently, some
benchmarkers offer project managers multiple "approved" options
such as Microsoft project or Prima Vera project software.

Standard: **The best practices project organization provides state-of-the-art soft-
ware, information, communication,
and project tools.**

Best practices project organizations use web pages, Notes based software, and email as primary global communications tools and to provide stakeholders access to important project information. Some organizations are reporting status of over 1,000 projects on web pages and Notes based software.

Information presented ranges from highly detailed to general in nature. Web pages and Notes based software are appropriate for transmission of information to globally dispersed individuals and sending alerts to teams. Several benchmarking organizations use web pages or Notes based software systems with folders or web pages set up for individual projects. Users can open a home page or folder that contains basic and summary information as well as status reports about each project. An index of projects and a brief description of each project can be provided.

Generally provided by project organizations is information such as project templates, methodologies, standards of performance, and best practices. Benchmark participants also report that their web pages contain background information for forthcoming briefings and meetings, training information and catalogs, competencies for grade levels, symposium materials, articles, and a link to the project management professional association (PMI) web site.

Web pages and Notes based software are suited for electronic project initiation, approvals, and status of requests. One benchmark organization reported that their global approval process was reduced from four months down to one week when they converted from a paper based approval process to an electronic one.

Benchmarkers caution that electronic communications are not without their drawbacks. Despite the advances of electronic reporting tools, many individuals and organizations continue to prefer paper based systems, particularly for more lengthy reports. In fact, the majority of organizations participating in the benchmarking forum still rely primarily upon paper for project charters, plans, and signature approvals.

Like other forms of communications, quality of information is more important than quantity and the method of transmission. Benchmarkers report that there is danger of information overload. Maintenance of the web pages and Notes folders is time consuming. Most benchmarking organizations report dedicating at least one person to developing and maintaining web based, email, and internet capabilities.

Other problems reported are that English based web pages and Notes software are difficult to use and understand in countries with different languages. In addition, there is often resistance to public dissemination of project information and sensitivity about everyone knowing the status of the project.

A constant challenge reported by benchmarkers is interfacing the various project related software packages with one another. For example, benchmarkers envision an ideal package that includes capability to easily construct graphical work breakdown structures and flow charts, develop schedules, be compatible with the host organization's accounting programs, and include budgeting, risk analysis, project tracking, and other miscellaneous functions.

10.2. NURTURE COMPETENCE THROUGH TRAINING

A vital component of the project organization's knowledge management role is training. Benefits can be immediately attained for the project and host organization by initiating competency based project management training programs. These can be internally generated or off-the-shelf and customized programs provided by universities and training organizations. Taught are the methodologies of the project management body of knowledge as well as other subjects such as leadership, international project management, working in political environments, and software proficiency. The most effective education programs focus on skills and best practices that can be immediately applied to the job. The result is a nearly instantaneous project management performance improvement.

As discussed earlier in this book, a major function of the Center of Excellence project organization structure is to raise project management competence across all company projects, facilities, and geographical locations. The process includes evaluating performance and comparing it with specific competency objectives and then building a compatible educational program. Project Support Office training tends to be specific to project related problems.

10.2.1   Benefits of Training

Emphasis on education is an area that differentiates the superior project organization. Research and benchmarkers' experience indicate that appropriate training immediately translates into improved project productivity and ability to achieve goals. Generated are high returns related to the investment. For example, sales techniques training for sales people can double revenues immediately after the instruction. In like manner, one benchmarking organization added 35% new people to their group in India. The new employees were given one month training. Their measurable project related productivity immediately increased far in excess to the cost of the training.

Several benchmark organizations state that training represents the *most valuable contribution* made by the project group to the host organization. Other benchmarkers whose organizations have no instructional programs, avow that training represents the *best potential way* to improve the organization and that a lack of training is their *most severe* problem. In the same vein they say that a major problem is being forced to work with inexperienced project managers who lack knowledge about the specifics of managing projects. Ingram's research of failed projects agreed. His investigation disclosed that project managers of failed projects repeatedly lacked competencies necessary to effectively lead the project. Specifically, people who were capable functional managers when assigned to projects often suffered a deficiency of entrepreneurial leadership skills as well as knowledge of best practices specific to project management.

Benchmarkers report that the need for training is particularly acute with people new to the project management discipline. A constant hurdle is bringing new people up to a high proficiency level in the shortest amount of time. Although experience is a superb teacher, it can also be slow and expensive when failures result. Benchmarkers say that the more efficient approach is through instructional programs.

> **Standard:** **The best practices project organization has a personalized development and training program based on identification of skills and competencies needed by the individual or group.**

10.2.2. Opposition to Training

Although the bottom line value of training is well documented, and benchmarkers unanimously agree upon its need, senior management in many benchmarking host organizations consider it an unnecessary expense. A survey of benchmarking organizations discloses that 50% offer no formal project management training. As a case in point, one of the largest organizations in the benchmarking group executes highly complex, long-term billion dollar plus projects, but they conduct no project management training. Their philosophy is that newly hired project managers should already be trained. Even for organizations that do conduct training, when budgets are cut, education is often one of the first sacrifices made.

10.2.3. Benchmarkers' Training Conviction

Despite the opposition from host organizations, benchmarking participants universally agree that training is a best practice of the superior project organization. In addition to the financial benefits previously discussed, the body of research literature and the experience of benchmarkers indicate that project management and leadership competencies are *learned skills*. People are not "natural born" project managers. They learn the actions and skills through experience, training, mentoring, and following examples of role models. Consequently, training is necessary in cases where project managers do not already possess the prerequisite skills.

10.2.4. Measure Training Results

Although benchmark forum participants agree about the value of training for project managers and team members, they also say that it must produce measurable and observable results. Forum respondents view training as an investment in better performance and recommend that educational activities focus on improving project manager best practices and competencies. Training programs and courses should result in immediate improvements that are observable and measurable. The results of the training can be evaluated through various methods including end-of-course surveys, tests for knowledge, post-training interviews, and performance measurements.

Standard: **The best practices project organization measures and quantifies the value of training.**

There is a wide variety of courses and seminars dealing with the skills of project management. Many hopeful students attend these courses. Benchmarkers attest that at the courses' conclusion, some students still can't effectively manage a project. Or worse yet, the approaches being taught are not correct and hinder rather than help the project team from achieving its goals. Consequently, one benefit of utilizing a competency based teaching approach is that it makes possible the correlation of teaching specifics with measurable project outcomes. Project management organizations should rightly ask consultants and academics to disclose the measured impact of their teaching activities on project goal achievement.

10.2.5. Types of Training Provided

Of the benchmarking organizations that conduct training, the majority (75%) utilize formal internal project management training programs. Outside training consists primarily of courses to prepare for the Project Management Institute's professional certification (48% of those conducting formal training) followed by paying for graduate level project management courses and undergraduate programs (36%) at universities and colleges.

Some of the benchmarking organizations have massive project management training programs. One is training approximately 17,000 managers in project management techniques with the ultimate objective of having most receive the PMP certification. Several are training over a thousand managers. Some benchmarking organizations have separate training units and facilities that they designate as their academy or university.

10.2.6. Evaluate Training Needs

Most benchmarking organizations start the training process with an evaluation of the skills and competencies needed by their project managers. The process is important because it establishes a baseline against which to measure the success of the training program and individual performance.

The highest and most immediately visible returns result when training is linked or applied to specific projects being executed by the trainee. One company reports that they use "just in time training." They go directly from problem discussion and resolution in the classroom to application in the workplace. For example, two large companies (one of which was a benchmarker) merged. They formed a project team whose objective was to combine the operations of the two companies. A project trainer was brought in. Each workshop session focused on the most immediate need of the team. The first class covered the development of a work breakdown structure for the project. Then was addressed the development of the merger project schedule using software. Other classes covered development of the merger budget, risk and political analysis, stakeholder communications, human resource plans, etc.

Organizations beginning the training process typically focus their efforts at lower levels of the organization. The logic is that the project group needs to begin developing a common vocabulary and approach. Subsequent training programs are usually linked to career path progress. Each of the jobs on the career path is defined and training is developed to optimize the individual's performance at each step up the ladder.

Sometimes training is a prerequisite to career advancement. A handful of benchmarking organizations require training and attainment of the PMP certification before advancing to higher levels of management.

10.2.7. Learning Mediums

There is general agreement among the academic profession that the *classroom* approach results in the highest learning levels. Students can make presentations and work in teams. Questions are answered on the spot and the flow of discussion is more immediate and fluid. Volume of material covered is higher than in other teaching mediums. The instructor can watch for visual feedback and signals.

On-line learning consists of conducting courses over the internet or intranet. One benchmark organization uses the on-line medium for project tutoring as well as for teaching specific courses. It is particularly suited for students who travel and for geographically dispersed project teams with limited access to the classroom environment.

Benchmarkers relate that often companies approach on-line learning as a way of reducing training expenses. Rarely does this turn out to be true when cost versus knowledge output is taken into account. Disadvantages of on-line learning are that it fails to offer students opportunity to make stand-up presentations and is less conducive to group activities. One benchmarking representative said that their on-line program was little more than a programmed learning approach with students responding to a checklist of activities and questions.

Flawless software and communications tools are mandatory for on-line courses. A large benchmarking organization set out to train several thousand project managers using a company intranet and internally generated software. A benchmarker who participated said that for a two week course, about one-half the participants were still trying to get on the system and into the class three or four days after its start.

Programmed distance learning is essentially a correspondence course. Each workshop consists of readings accompanied by a series of activities. Correspondence type learning is disappearing in educational organizations as well as most companies.

10.2.8. Types of Training Offered

Most companies in the benchmarking forum who provide training offer five to seven project management specific courses to project managers and project team members. Most tie their training approach to the philosophy espoused by the Project Management Institute as described in the *Project Management Body of Knowledge Guide*. Introductory overview courses are offered as a minimum. Several benchmark organizations have specific performance development programs for each level of management. Most train toward a company proficiency certification or certifications provided by outside training organizations or the Project Management Institute's PMP professional certification.

Basic training programs surveyed cover project management methodologies, tools, and techniques of managing projects. Also emphasized are techniques of leadership and managing people. Some companies that employ predominately technical people feel that leadership and people management training is so important that as many as three courses out of seven transfer this type of knowledge.

Companies that are globally diverse offer courses in international project management. These programs are judged essential because they cover topics not normally addressed in other project management courses, but which can have strong impact on the project leader's success. Examples of such topics are the global and cultural impacts of religion, politics, race, manners, dress, appearance, mannerisms, speech, and language.

Numerous other training courses are provided depending upon the needs of the project organization. More recent courses developed address effective electronic written communication and working in virtual project team environments. Others provide instruction in software and tools usage, contracts basics, and project finance. Some deal with human skills such as managing project sponsors and working within the realities of the organizational environment, politics, and culture.

10.2.9. Course Durations

Duration of the courses varies. Several organizations offer an "Introduction to Project Management" course for executives that is two hours in duration. Also sponsored are four hour "Why Project Management?" courses. Numerous other courses are offered in two to five day versions. Often the introduction and overview of project management is a five day course.

10.2.10. Promote Life-Long Learning

Forum attendees profess the need for maintenance of project management skills, acquisition of new skills and developments in the profession, and knowledge about the latest tools and software. This is accomplished through continuing education. One participant's company requires 40 hours of training each year. The majority of benchmarking organizations consider the maintenance of skills the responsibility of the individual. Continuing education is received from classes and seminars and attending professional conventions. Most organizations in the benchmarking forum provide support through funding.

Standard: **The best practices project organization commits to life-long learning for project managers and teams.**

10.2.11. Reward Training

For training to be most effective, benchmarkers say there should be tangible rewards, recognition, and accolades. As a minimum they might consist of receiving certificates for courses completed and having the accomplishment recorded in the employee's personnel files. Others encourage praise from management for completing educational courses. For completion of longer duration programs such as passing the PMP certification exam, rewards might consist of financial bonuses, qualification for promotions, and formal recognition during yearly performance reviews. A few benchmark organizations give a 5% pay increase for completing a seven-course certification program and a 10% raise for the PMP. Another gives a $5,000 bonus for passing the PMP.

References: Ingram, 1998; Kirkpatrick and Locke, 1991.

10.3. MANAGE KNOWLEDGE BY ENCOURAGING EXCELLENCE

Leadership in setting and building high standards of excellence is a responsibility of the project organization. Established are best practices of superior project managers and organizations. Resources from outside the project organization are utilized to validate its efforts and to provide benchmarks upon which to model improvements. Provided are ways to enrich individual project managers and projects through by patterning behavior after that of role models and mentoring.

10.3.1. Set High Standards of Excellence and Quality

Superior project organizations build excellence and quality into all activities performed. An "excellence" or "quality" state of mind is created where *superior* performance is the accepted role rather than *acceptable* performance. Over a period of time excellence and quality become institutionalized and part of the organizational culture.

Standard: **The best practices project organization builds excellence and high quality into every activity performed.**

One-third of benchmarking participants indicate that they formally plan quality or excellence into each project plan and each phase of the project process. The project organization itself is reviewed to ensure that quality and excellence are integral ingredients in the overall formula for success.

Superior project organizations build excellence by defining the global essentials or best practice standards that project managers and groups aspire to attain and maintain. The primary objective of this book, its companion volume, *The Superior Project Manager*, as well as all the activities of the benchmarking forums, is to assist project executives and managers in accomplishing this goal. The two volumes define the best practices and standards of superior project performance. They serve as a baseline of superior performance and represent core values that are to be applied in a broad and flexible manner.

As used in this book, the terms "standards," "best practices," and "competencies" are used interchangeably. They are the overarching rules, principles, practices, models, and measures established by the participants in the benchmarking forum. Standards and best practices are regularly and widely used and familiar among superior organizations in the project management field. They are substantially uniform and well established by usage in the speech and writing of the educated within the field. They establish benchmarks of superior performance.

Standards and best practices have several distinguishing characteristics. They tend to define what a project manager or organization *should do* rather than what they *should know*. They characterize the manner in which the project manager or organization performs the job rather than being reflective of the success of the project. Best practices and standards are observable and measurable and drive superior rather than average performance. They have been scientifically validated and are weighted in terms of their importance in increasing the probability of project goal achievement. They should consistently predict and distinguish superior from average performance.

10.3.2. Use Resources and Knowledge from Outside the Project Organization

Benchmarkers report that often the best way to improve the project organization is to bring in outside references and resources. They say that the adage, "You're never a prophet in your own village" is often true in large functional organizations. Sometimes concepts are more readily accepted as valid when they originate outside the organization.

There is logic and research support for the belief. Studies of decisionmaking techniques conclude that it is easier to simply copy and improve upon best practices and experiences of others than to reinvent the wheel. Best practices organizations benchmark and form alliances with educational and professional organizations. The result is that the probability of success is higher. More importantly, the organization is more rapidly raised to higher standards of excellence and performance.

Standard: **The best practices project organization benchmarks and implements best practices from other individuals, teams, and organizations.**

10.3.2.1. Benchmark with Best Practice Organizations

By interacting with other project organizations, project groups gain credibility and acquire access to a database of best practices. These serve as ammunition and strengthen the group's knowledge base.

Benchmarking also serves as a source for continuous improvement since best practices organizations are constantly evaluating their performance and state of the art in comparison with other project organizations. Professor Nutt found that benchmarking was the most successful way of making major decisions. His research concluded that two years after benchmarking based decisions were made, they were successful in 96% of the cases analyzed.

Benchmarking is a basically a two-step process. It involves, first, observing other organizations and identifying best practices. The second step is that those practices are copied and implemented. Benchmarking varies in the specifics of its approach. Some are data intense and include comparisons of project process and performance data. Others deal with broad conceptual issues where project managers and stakeholders discuss and search for best practices relating to specific problem areas or topics. The most successful benchmarking decisions in Nutt's studies of decisionmaking methods were made when a problem was identified and a search conducted for solutions. Then the solution was implemented. In numerous cases benchmarking involved observing the practices of several different organizations and amalgamating the best features from each.

Benchmarking encourages comparing and contrasting one's own organization with others and offers several benefits for the organization seeking excellence. From a cost view, benchmarking is one of the least expensive ways of making decisions as other organizations have already performed the pioneering work. Benchmarking opens the group to new ideas and bypasses the "not invented here" syndrome. Using the benchmarked organizations as examples and standards of performance provides credibility and validation to the quality of the resultant decision. Entry is also provided to a library of data, performance, and materials that support the search for excellence. Benchmarking results in membership in a network of other superior performers who are generally eager to advance the profession and are willing to share helpful information.

Reference: Nutt, 1999.

10.3.2.2. Partner with Educational and Professional Organizations

Associations outside the organization tend to lend an aura of professionalism and believability. Organizations that have partnered with educational and professional organizations find that the standards of project group performance are invariably raised.

10.3.3. Be a Role Model of Excellence

One of the best ways for individuals to make correct decisions and choices is to benchmark the performance of respected individuals. In other words, to copy, tailor, and pattern one's behavior after the successful approach taken by a superior project manager role model. The process works well in project organizations because it gets best results when it is informally and constantly put into practice.

The validity of the concept applies to teams and groups as well as individuals. For those initiating a project organization, a good starting point is to use internal teams that already do projects well. Even though not fully trained, these teams take the lead and serve as role models.

Project organizations can generate excellence by serving as role models for the project managers and teams. According to the research of Toney and others, business leaders who take their role model position more seriously consistently have higher performance than leaders with lower levels of role model consciousness. Being a role model means that the project organizational leaders apply frequent actions to guide and shape the behavior of subordinates. Rather than words, the project organization's actions exhibit and validate their beliefs and attitudes about team performance, expected behavior, and best practices. When serving as a role model, the project organization's example sets the overall tone for project speed, efficiency, and effectiveness.

Standard: **The best practices project organization acts as a role model to communicate positive values and guide the behavior and performance of the team.**

Researchers and students of leadership from the earliest times recognized the significance of leading by example. Research pioneer Fayol declared that the leader should set a good example and generate energy, initiative, and loyalty among the personnel. He said, "One of the most effective methods of training is by example." He also stressed that setting a bad example results in serious repercussions on the unit as a whole. Researchers Kirkpatrick and Locke asserted that effective leaders present themselves as role models and personally act in a way that is consistent with their vision. As a result, followers are motivated to attain the vision and to perform in a superior fashion. The importance of serving as a role model is apparent in times of crisis, stress, and discouragement, concludes Avolio's investigation. The projected self-confidence when confronted with unusual events, gives rise to leaders being viewed as charismatic, if not inspirational.

References: Nutt, 1999; Toney, 1996, 1998; Avolio et al., 1993; Kirkpatrick and Locke, 1991; Fayol, 1916.

10.3.4. Mentor for Excellence

The importance of mentors in attaining excellence is well validated by the research. In general, the person being mentored will rise in capability to the level of the mentor. Tests in production environments have found that new employees when teamed with the worst worker in the group will perform at the level of the worst worker. If the same person is placed with the best employee, their performance rises to the best employee's level. When McDonalds Corporation hires teenagers they are paired with the very best.

A *personal* mentor is an individual that assists another individual. They tend to be respected role models and provide coaching and even counseling. The relationship is more personal than that described in the previous paragraph where the mentor simply helps another learn a job. Often personal mentoring relationships are long-term and may even span a lifetime.

Benchmarkers say that the personal mentoring concept is superb but difficult to implement. One reason is that the individual most people would select as their mentor is usually a senior or highly respected person who is excessively busy. As researcher Nutt determined, the best employees are also the best mentors. These are also the individuals that are most difficult to remove from their regular job. In addition, in most political and cultural settings, often the potential mentor represents a competitor for future promotions. The end result is that almost none of the benchmarking organizations reported having a *successful* personal mentoring program. The few that were tended to obtain mentors from outside the organization.

Standard: **The best practices project organizations take a proactive stance in developing programs aimed at attracting, selecting, retaining, and rewarding top performing project managers.**

10.3.5. Set a High Standard of Truthfulness

Serving as a role model and setting high standards for the application of truthfulness and honesty have a direct impact on project success and efficiency. Truthful organizations, teams, and individuals are consistently more successful at achieving goals than the unethical. Honesty is essential to project leadership competence.

Standard: **The best practices project organization is truthful in all dealings and relationships.**

Modern business scientists agree that truthfulness is a core competency essential to goal achievement. On a broader scale, Lewicki's 1998 research established that truthfulness and its associated outcome, trust, are *mandatory* for the maintenance of social and organizational order. The Toney study of Nasdaq chief executive officers shows that the single factor with the highest correlation to company profitability is truthfulness. Kouzes and Posner's 1993 survey of 15,000 people finds that credibility is the single most important attribute of leaders. Honesty is judged so important by the Academy of Management that the entire July 1998 issue of the research journal *The Academy of Management Review* is devoted to the topics of truthfulness and trust.

10.3.5.1. Benefits of Truthfulness and Trust

The results of trust can be measured in the form of increased sales and reduced costs as well as reduced risk and a myriad of general organizational and personal benefits.

10.3.5.1.1. More Sales. People buy from those they trust. Project teams with the reputation for honesty have a natural advantage in the sales arena. Confidence is greater that the project team and any associated products or services will perform as represented. After the project agreement is reached or the project related product or service is purchased, there is increased assurance that performance will be as characterized. When problems arise, there is more conviction they will be resolved and the product or service will be supported.

Speed to market is also increased. According to the research of Lewicki, trust improves organizational sales and marketing efficiency. A high level of trust can be translated into faster performance. When applied to new product development and other projects, increased speed to market means more sales, higher profits for the first market entrants, and improved market share.

10.3.5.1.2. Lower Costs and Better Efficiency. Truthful project teams record a myriad of advantages in the area of lower costs and greater efficiency.

- **Increases team output and efficiency.** Controlled tests of small groups conclude that project teams composed of members who have a high degree of trust are more efficient and achieve greater output. The trustful groups spend more time on productive activities and less time on paperwork, writing agreements, and building control mechanisms. When events occur that insinuate someone might have done something questionable, the truthful group assumes a positive yet time-efficient stance, i.e., that the person did not intend to do a wrongful act.

  Rousseau's 1998 data support that trust is an important element for giving team members the faith to self-organize. Helpfulness, cooperation, and a feeling of responsibility for the group are descriptors of cohesive team relationships. Truth and trust foster the subjugation of personal needs and ego for the greater common good.

  Truthfulness makes complex strategic alliances possible. Success is rarely a reflection of the contract that defines the terms of the joint effort since it is impossible to monitor and provide controls for every detail. Successful alliances are characterized as honest relationships with commitment from all partners and confidence that each partner will perform as represented.

- **Behavioral consistency and predictability.** Consistent behavior over time and in various events gives team members the ability to predict courses of action that are consistent with leadership strategy and philosophy. Team members become more willing to accept the risk of taking action without specific guidance from project leadership.

- **Reduces need for control mechanisms.** Multiple studies show that trust is a substitute for control mechanisms. Control mechanisms are created when adequate trust does not exist. They are expensive and time consuming to develop, monitor, and enforce. Note that control procedures and tools do promote cooperation where limited or conditional trust exists.

In distrust, the individuals begin establishing rules of behavior and establishing control devices. Various legal, personal, and institutional mechanisms are used to control mistrust, such as

- **Contracts.** Contracts become more lengthy, detailed, restrictive, and specific as distrust increases.
- **Auditors.** Distrust results in more auditors, quality control inspectors, and sanctions for infractions.
- **Due diligence investigations.** Background research, credit reports, formal planning, monitoring, and reporting guidelines all reflect the degree of trust.
- **Deterrents and structural safeguards.** Include costly reporting and checking devices, written notices of departure from contracts and agreement, audits, cost control, quality control, arbitration clauses, and lawsuit provisions.
- **Regulations.** Define the expectations for behavior. They also add cost to the organization. They take the form of laws, standards by professional bodies, and contractual constraints and guidelines.
- **Guarantees and warranties.** Protect against the failure to perform.
- **Information restrictions.** Access to confidential information might be limited.
- **Threat of disclosure.** If an individual or organization cheats on a credit application, they face the risk of disclosure. Future credit ratings will be negatively impacted individual.

- **Increases communications accuracy and openness.** Truthfulness and trust make possible the free exchange of knowledge and information. Trust is necessary before sensitive materials and information can be divulged to others. Knowledge is a source of power. Turning it over to others increases the risk that it could be misused. Employees perceive as truthful managers who take the time to explain decisions and answer sensitive and embarrassing questions in a forthright manner.
- **Improves flexibility.** With trust there is faster response and

more flexibility to adjust to new situations not covered in the original contract. Trust is manifested by flexibility. Incomplete contracts or phased contracts are made possible only when there is a high degree of trust. Many information systems projects are based on technology that is changing as the project unfolds, and the customer desires the new technology. Some projects are being bid in phases. Trust ensures that neither party will be taken advantage of as the scope of the project unfolds and changes.

- **Trust is self-curing or self-healing.** Trusting relationships require constant monitoring, quick correction of inaccuracies, and repair of trust as tensions arise. Trusting individuals promote interdependence and cooperation. McKnight's 1998 studies disclose that in trusting relationships, evidence that appears to contradict the trust is viewed as suspect and is denied or considered unimportant. As a result, participants do not waste time attempting to determine motives.

- **Encourages help seeking behavior.** There are often constraints that discourage individuals from seeking help from their colleagues. There could be a fear that others would think them inadequate or unqualified for the job.

- **Ensures fairness.** Trust assumes fairness and equity with each party confident they will be rewarded in relation to the amount contributed.

- **Fosters confidentiality.** Truthfulness assures people that their interests will be protected as well as promoted, they can feel confident about divulging negative personal information, they are assured a frank and full sharing of information, and they are willing to overlook and understand apparent breaches of the trust relationship.

- **Encourages empowerment and delegation of authority and control.** A key outcome of truthfulness and trust is the willingness to give subordinates authority and control. Empowerment and delegation reduce costs by having decisions made at the lowest and most economical level. Trusting is also directly related to the capability to manage larger spans of employee control. Funds expended on supervision are reduced.

- **Reduces legal fees.** One of the most visible measures of truthfulness and ethics is the amount of funds expended on legal fees. Those who are distrustful spend more on litigation and defending accusations. Other penalties directly attached to lying may be legal and court imposed. The failure to keep a commitment may involve sanctions, including paying the costs of compliance.

- **Being truthful and trusting is the easiest approach.** Investigation by Gareth and George supports the conclusion that in new relationships, trust is the easiest, most efficient, and effective option. To determine if the other party isn't worthy of trust requires considerable effort and time. Even so, initially the trust is typically conditional and controls are usually in place. Trust is extended as long as the other party behaves appropriately, reciprocates the trust, and conforms to the controls.

10.3.5.1.3. Trust Reduces Project Risk. Projects are conducted in an environment of uncertainty. Bhattacharya in his 1998 empirical studies concludes that trust reduces risk by increasing the predictability or expectancy of the behavior of individuals and teams. As a result, trust augments the willingness of the participants in the project to take short- and long-term risks. Since trust is directly related to risk and the reduction of risk, the project goal achievement aspects of trusting and truthfulness are solidly linked to the fundamental objective of maximizing project performance while minimizing risk.

- **Dependency and vulnerability related to risk.** Trust involves placing one own's and the project team's fate, dependence, or vulnerability in the hands of another according to the 1998 research results of Sheppard and Sherman. Trust is a particularly relevant factor to project goal achievement any time there is potential for one or more of the participants or stakeholders to lie, exit, betray, or defect. There is never a guarantee that the trust given by the first party will be reciprocated. Risk is reduced because the trustful team member or project manager will perform as they have represented. As a result, action is taken based upon the representations of the other party.
- **Trust creates more trust and reduces risk.** Initially, the parties involved tend to take small risks and limit their commitment. As positive experience accumulates, it serves as justification for increased risk taking and investment in the association. Each step along the way is measured by increasing amounts of trust.
- **Risk is reduced over time.** Doney's 1998 research shows

that the trust/risk relationship is reciprocal and builds upon itself. When the trustee knows that the trustor is taking risk based upon truthfulness, there is a tendency to continue to behave in a truthful manner. The reciprocal nature of the relationship is a key element in building and strengthening the relationship. The risk element is mandatory because truthfulness is demonstrated only when the other party takes action based upon the truth. Trust emerges as the parties determine that past actions provide a foundation upon which to predict positive future performance. Trust building ideally includes information about the individual's past actions. Sources come from experience, word-of-mouth, or reputation.

10.3.5.1.4. General Efficiency Improving Benefits.

- **Serves as evidence of other values.** Truthfulness serves as evidence of a broader value set in the truthful person. The participants assume that truthfulness is the cornerstone in a foundation of other solid ethical competencies and best practices.

  Truthfulness tends to maintain ethical focus on the part of the truthful individual. Inversely, lies blur ethical focus. Teams that suffer from dishonesty tend to exhibit greater amounts of goal incongruence, negative internal politics, shifting coalitions, and conflicts, concluded Elangovan and Shapiro.

- **Higher quality team members.** People prefer to work with project managers and project team members that they admire and trust. Consequently, the pool of prospective team members is larger and the quality is greater.

- **The value of truthfulness is geometric.** The experience of the truthful individual with others has a multiplier effect with the result that the person gradually acquires an expanding reputation for honesty, documents McKnight in his research. Testimonials and word-of-mouth certify as "proof sources" that the person can be trusted. The parties remember the behavior, accumulate it over time, and communicate the experience with others. Reputation has high value when isolated negative events occur.

  Lies too are geometric. However, lies have disastrous consequences because each lie requires additional lies to support it. Eventually, the weight of the untruths and the support needed to sustain their logic becomes overwhelming and collapse occurs.

- **Reduces stress and anxiety.** A high degree of trust results in less individual and organizational anxiety and stress, according to investigator Bhattacharya. Working conditions are improved, morale is higher, and medical expenses are reduced.
- **The truth builds stronger relationships.** Trusting relationships with other people, the project team, and organizations are more solid; they tend to endure longer, concluded the 1998 research of Sheppard and Sherman. As truthful experiences multiply, the relationship between the parties becomes increasingly interdependent and more tightly bonded together.

  The relationship between the trustor and the trustee is an important factor in influencing the potential for lying. It's important to note that people are in a position to betray *only* because they have established themselves as trustworthy. When trust is betrayed, the result is a strong set of negative emotional responses, detailed researchers Gareth and George in their 1998 work.

10.3.5.1.5.　Other Benefits.　Many benefits of trust are personal. Trust affects one's reputation, and perception of an individual's degree of trustfulness can have a powerful effect on a career. Other benefits are non economic in value. Examples are friendship, support, and social approval. Trust is built on expectations, which are partially emotional.

10.3.5.2.　Building Trust in Project Organizations

- **Tell the truth.** The fundamental best way to build trust is to implement the project organization best practice, "Be truthful in all dealings and relationships."
- **Establish controls.** As the accounting profession has known for years, one way to build trust and security is to serve notice that honesty is being monitored. Establishing controls does this. The degree of detail depends on the situation and the amount of risk involved. For example, more controls are justified when the project manager is in a legal and fiduciary role, e.g., managing the funds of others. On the other hand, those who have partnered in trustful relationships for years may have virtually no formal controls.
- **Structure the organization to encourage trust.** The organization can be structured to encourage truthfulness and trust. Policies and team culture can teach and reinforce high

standards of honesty in relationships among employees, project management, and other stakeholders.

- **Communicate.** Open and prompt communication is indispensable. The reciprocal process creates credibility, a sense of openness, and intimacy.
- **Reward trust and truthfulness.** Patterns of honesty should be rewarded. For example, one company has a formal evaluation process that rewards project teams who develop a track record of truthful dealings with increasingly larger project responsibility.
- **Punish dishonesty.** The threat of sanctions and/or the *disclosure* of dishonesty serve as an effective deterrent, concluded Hagan and Choe in 1998. For the unethical party, disclosure could prove economically devastating if stakeholder associates learn of the dishonesty.

  People lie when they perceive a low likelihood of suffering severe consequences. Both the likelihood and the severity of the penalty are analyzed before action is taken. Also analyzed is the possibility of being identified as the perpetrator of deceit.

  Elangovan and Shapiro add that expectation of forgiveness is also a factor. In cases where the chance of detection is high, there could also be a high expectation of forgiveness. Forgiveness essentially means that there will be no penalty attached to the betrayal.

References: Bhattacharya et al., 1998; Doney et al., 1998; Hagan and Choe, 1998; Ingram, 1998; Elangovan and Shapiro, 1998; Lewicki et al., 1998; McKnight et al., 1998; Rousseau et al., 1998; Sheppard and Sherman, 1998; Whitener et al., 1998; Kramer and Tyler, 1996; Barney and Hansen, 1994; Rogers, 1994; Kirkpatrick and Locke, 1991; Bass and Avolio 1990.

11

Supporting Project Manager Performance

As has been stated numerous times throughout this book, the most important single factor impacting project success is the project manager. Benchmarkers confirm that the project manager is the hub around which all project organization activities revolve. This single individual is responsible for between 35 and 75% of the project's success. The project manager is the person who maximizes the team's performance. They are the organizational representative who can be sent to remote locations and depended upon to capitalize on controllable events and mitigate the uncontrollable. When projects are in trouble a first step in resolving the problem is to send in a superior project manager.

If the project manager is viewed solely as a financial asset that represents a sizeable organizational investment, the potential for revenue generation is significant. When the total cost of hiring, training, building experience, mentoring, and imprinting loyalty through indoctrination in the organizational culture are added to the potential costs of firing and rehiring, the total investment in the superior project manager is remarkable.

As a consequence of the importance of this individual, it seems almost without question that one of the most important roles of the project organization is to provide support for the project managers and their teams. One way the support is manifested is through programs to attract, hire, motivate, reward, and retain the highest competency level project managers with potential for long-term growth.

The project organization also provides support by serving as an ambassador in assisting and representing the project teams in cultural and political situations.

11.1. PROVIDE SUPPORTIVE LEADERSHIP

The prevailing approach adopted by the project organization in governing project managers is supportive, facilitative, or servant leadership. The three terms are used interchangeably. Facilitative leadership encompasses the philosophy that the role of the project organization is to support and help project managers and their teams. For supportive leadership to work one must build the project organization with skilled and competent project managers, delegating to them authority including the trust to perform without authoritarian interference and encouraging broad spans of control.

The research of James Worthy at Sears showed bottom line benefits from what he labeled servant (supportive) leadership. In stores where management demonstrated trust, support was evidenced by broad spans of control and extensive delegation, higher sales, better profit performance, and lower payrolls as a percentage of sales. Employee morale was higher, more people were promoted to positions of authority in the company, and caliber of people was judged better. Quality of planning was considered better because it was developed at lower levels from people closer to the front lines. The impact on the bottom line was positive because expensive management was reduced and decisions were pushed to lower and more economical levels. Indirectly, sales were increased because morale and enthusiasm were higher.

Standard: **The best practices project organization provides supportive, facilitative, and servant leadership for its project managers and teams.**

A prerequisite for successful supportive management and leadership is highly qualified people. Consequently, the project organization has a strong incentive to seek out, nurture, and retain superior project managers. The desired capabilities of project organization managers tend to reflect those of Worthy's "servant" leader; they willingly assume responsibility, are trustworthy, use good judgment, and can work on their own without constant guidance.

The danger of the supportive approach to management is the resultant increased authority of project managers tend to amplify the risk of mistakes being made. The trusting facilitative project organization encourages project managers to use those mistakes as learning tools. Overall, the nature of trusting supportive project organization leadership is that it *forces* project managers to exercise initiative, develop self-reliance and confidence, and to learn from errors and experience.

The supportive attitude is by definition *people* oriented as compared to *methodology and tools* oriented. Its evidence is numerous people oriented best practices. It includes providing project managers and teams with the information needed for job performance. The supportive project organization places faith in the ability of project managers to maximize opportunities and to minimize problems. Once the supportive approach to management is put in place, problems tend to take care of themselves. For example, if the performance of a project is suffering, the supportive approach would search for ways the project manager and team could be assisted and provided resources for improvement. Solutions would be management strengthening rather than methodology improvement oriented.

Supportive management emphasizes team building, training, coaching, and mentoring. The objective is to improve the levels of confidence of the project managers and teams so that they can take on even greater amounts of authority and responsibility while minimizing their need for supervision and direction.

References: Worth, 1952. (See *The Journal of Leadership Studies,* 1998, Vol. 5, No. 4, for a complete issue dealing with subsequent research and best practices associated with Worthy's research.)

11.2   HIRING THE SUPERIOR PROJECT MANAGER

Benchmarking participants say that the strength and success of the project organization ultimately rests upon its ability to staff itself with superior project managers. Hiring promising project managers is the first step in building the best practices project organization. One company reports that project manager and team selection is considered so important that their project portfolio is developed around the availability of project managers rather than other resources such as funds. They judge that the professional and qualified project manager is their most scarce resource and major constraint on strategic success.

Standard: **The best practices project organization hires superior project managers and those with potential to be superior project managers.**

Benchmarkers report that they search for three broad competency groupings when evaluating prospective project managers. They are (a) character, traits, and background, (b) professionalism consisting of leadership and management skills, and (c) project specific skills comprised of the application of structured methodologies and procedures.

Figure 11.1 presents a portrait of the superior project manager and associated validated competencies that are supported by empirical research. Each of the competencies is discussed in detail in the companion volume, *The Superior Project Manager.*

References: Ingram, 1998; Kirkpatrick and Locke, 1991; Lieberson and O'Connor, 1972; Likert, 1961.

11.2.1. Character, Background, and Traits

11.2.1.1. Character

Character anchors the superior project manager. Of all the factors that describe character, the most important that correlates with goal achievement is truthfulness (honesty). Truthfulness is essential to leadership competence. It serves as the foundation upon which goal achieving behavior is based and trusting relationships are formed. Honesty is so important that it compensates for a lack of leadership expertise in other areas. As the father of modern management, Henri Fayol stated in his 1916 book, *General and Industrial Management,*

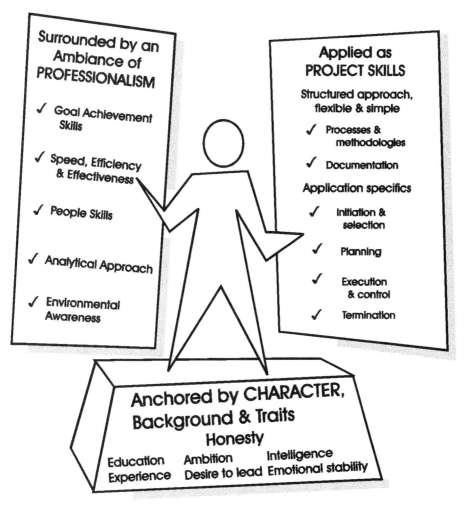

THE SUPERIOR PROJECT MANAGER

Surrounded by an Ambiance of PROFESSIONALISM

✓ Goal Achievement Skills

✓ Speed, Efficiency & Effectiveness

✓ People Skills

✓ Analytical Approach

✓ Environmental Awareness

Applied as PROJECT SKILLS

Structured approach, flexible & simple

✓ Processes & methodologies

✓ Documentation

Application specifics

✓ Initiation & selection

✓ Planning

✓ Execution & control

✓ Termination

Anchored by CHARACTER, Background & Traits
Honesty
Education   Ambition   Intelligence
Experience   Desire to lead   Emotional stability

FIGURE 11.1

"The slightest moral flaw on the part of a higher manager can lead to the most serious consequences."

11.2.1.2.   Background

Background impacting project goal achievement is education in the form of a college degree and/or PMP (Project Management Professional) certification. At least two-and one-half years' experience managing and leading teams is also a key factor in successful management of projects and leading groups.

Although the specific makeup of a project manager's background is not a precursor to success, it can have a significant influence on the ability to achieve goals. The most important element of the superior project manager's background is experience with activities related to goal achievement, leadership, management, and working with teams and organizations. Superior project managers typically have college degrees in business or engineering and professional project training and certification. Functional experience in the project specific technical area, work experience, education, and experience with the project team all have a positive impact on the bottom line. What seems to have little impact on project management performance is age.

11.2.1.2.1. Education. Project organizations in the benchmarking forum recommend that prospective candidates for the project manager position will ideally possess bachelor's degrees and be PMP certified.

- **College or university degree.** Studies agree that a business or engineering degree is an asset. Seventy-five percent of benchmark organizations require a bachelor's degree for its project managers. In a more general study of top performing business leaders, researcher Chandy determined that 57% had business or technical undergraduate degrees. Sixty-six percent have at least one graduate degree and of these 60% were MBAs.

References: Chandy, 1991.

- **PMP certification.** Less than 10% of forum participating organizations currently require PMP certification of its top project managers. Even so, the consensus of members of the Top 500 Project Management Benchmarking Forum is that the PMP certification from the Project Management Institute is an indicator that the project manager has a foundation of knowledge needed to successfully manage projects. As a consequence, it is judged a best practice of superior organizations.

  Like a college degree, the Project Management Professional certification indicates that the holder understands the basics of project management and has approached the subject as a professional discipline. It also establishes a common vocabulary and instills in the holder the value of core activities such as the work breakdown structure, the project schedule, budget, and risk analysis. The certificate is held in high regard by people outside the project management profession, is globally recognized as reflective of a predictable and universal approach, and signifies that the holder has a minimum level of experience and knowledge about project management. Benchmark organizations that staff with PMPs say that the practice lends the appearance of professionalism and can be effectively promoted when dealing with potential clients.

Standard: **The best practices project organization hires project managers who possess the Project Management Professional certification.**

Criticism of the certification process is that it has little correlation with project manager ability to execute a project. Although experience is a requirement for certification there is little in the test to indicate the test taker can develop a project charter, then apply a structured methodology that results in a project plan, procedures for execution and control of the project, and termination. Benchmarkers complain that sometimes they hire people who are PMP certified only to find that they are incapable of putting a project together and managing it.

11.2.1.2.2. Project Leadership and Management Experience. The candidate's leadership and management work experience is a significant factor in predicting future project governance success. Empirical support can be shown for links between prior leadership and management working experience among the project team members and project success.

It's important to note that two studies have shown that the duration of the experience seems to be immaterial after approximately two-and-one-half years. In other words, the work performance of the individual seems to improve for the first two-and-one-half years on any job and then levels out.

References: Norburn, 1989; Govindarajan, 1989; Hise and McDaniel, 1988.

11.2.1.2.3. Functional Training in the Project's Subject Area.
The functional training in the project specific subject area is positively correlated with competence and ability, although the correlation is less than that for general leader and managerial experience. Project teams that are most likely to achieve superior performance are those that match the requirements of their technology, project management, and leadership strategies with the background and expertise of their project managers.

References: Chandy, 1991; Braham, 1991; Bassiry and Dekmejian, 1990; Govindarajan, 1989; Norburn, 1989; Miller et al., 1985; Gupta and Govindarajan, 1984; Hambrick and Mason, 1984; Hitt et al., 1982; Kotter, 1982.

11.2.1.2.4. Has Learned from Failure.
Bibeault states that the phenomenal success of many superior leaders has followed on the heels of an earlier failure. It is his conclusion that the lessons learned in failure moderate the personal character and abilities of the successful leader. The process involved when a project manager faces adversity and impending failure is described as similar to tempering to remove impurities from steel. It's a violent process but the end product is more refined, durable, and capable of holding a sharper edge.

Standard: **The best practices project organization values lessons learned from prior failures in project manager candidates.**

Scientific support for these opinions is broad. Leaders consistently avow that the impact of previous personal failure as a learning tool is significant, states Toney's 1996 studies. They suggest that the initial learning curve in the first job as project leader is steep. These failures are not reflected in most studies of leaders because they occur in the first year or two of becoming a project leader, or early in the prior experience of currently successful leaders.

In a 1990 study, Makridakis found that leaders who have *not* experienced major failures tend to be biased in their decisionmaking processes. Since they do not have a clear understanding of the factors that cause failure, they are not conditioned to recognize signs of impending disaster and to take preventative action.

References: Doyle, 1994; Fagan et al., 1994; Cooper, 1979; Bass, 1961.

11.2.1.3. Traits

Other traits of less importance are ambition and a desire to lead, a service attitude or gaining satisfaction from helping others, above average intelligence, self-confidence, faith in a positive outcome, and emotional stability.

11.2.2. Leadership and Management Skills

After the foundation of character, background, and traits is set, the superior project manager is surrounded with an ambiance of professionalism. Professionalism includes all the knowledge and skills attained in typical college business and engineering courses. Professionalism is exhibited through application of the skills related to leading and managing groups of people. A moderately high degree of competence in the project's technical field is also a factor. An overarching skill is the ability to interact with and motivate people.

The probability of project goal achievement is significantly increased when the project manager is able to paint a vision, maintain constant focus on goals, and tie the project's strategy to that of the host organization. The most successful project managers critically analyze alternatives and opportunities. They seek the right answer rather than building support for a preconceived idea. They scan the environment and are aware of environmental events, threats, and opportunities. They optimize all channels of information.

11.2.3. Project Skills

Even with the correct character, background, and professional skills, the superior leader must be proficient at the technology of their field. In the world of project management the superior project manager displays a high degree of expertise in the application of structured project goal achievement methodologies and procedures. The methodology is broad based and flexible and the most simple that is appropriate for the project being managed. The superior project manager thoroughly understands how the methodology of project selection, planning, implementation, and termination are applied effectively and efficiently to a variety of projects in diverse global and cultural environments. They fully comprehend how the application of character and professional leadership and management skills optimize team performance during each project phase.

Note that a deficiency in project specific skills might not be critical to the hiring decision. There is a general feeling among benchmarkers that project specific skills can be rapidly taught if the student possesses the necessary character and professional skills.

Benchmarkers say the best way to evaluate the candidate's knowledge of project specific skills is through a series of verbal or written questions. For example, the interviewer can ask such questions as, "What are the three most important things you do when running a project" or "What would you do in the following situation?" To determine if the candidate understands how to plan and execute the project, they can be asked to describe how they would develop plans to build a house and then execute the project. The answer to any of these questions should evidence that the candidate knows how to develop a structured methodology. The interviewer looks or listens for key words (such as work breakdown structure, project plan, budget, risk analysis, communications plan, variance analysis, and termination plan) that indicate the candidate is applying a structured approach or methodology and is familiar with professional project management terminology.

11.3. PROBLEMS IN PROJECT MANAGER COMPENSATION AND MOTIVATION

Attention of benchmarking organizations has focused on project manager compensation, motivation, and career tracking for a couple of reasons. The first is a shortage of professional project managers. Organizations increasingly recognize the value of the superior project manager as a revenue generating and cost saving individual. This increases demand for the existing pool of project managers. It is natural that salaries rise as a response to the need to remain competitive.

Second, benchmarkers say that project manager pay is in a state of transition because of the increased importance of project organizations and the management of multi-functional projects in large enterprises. Historically, most functional organizations developed their compensation, motivation, and career paths around traditional functional units or departments. Now they recognize that reward and motivational systems need to reflect the expanded role and multifunctional nature of the project manager position. As a consequence, best practices organizations are reviewing their compensation programs and developing appropriate ways to reward, motivate, and retain project managers. The process is faced with numerous challenges.

11.3.1. Project Manager Positions Are Unclear

Historically in large functional organizations, there are hundreds and sometimes thousands of individuals with the project manager job title. The widespread use of "project manager" as a generic title makes it difficult to differentiate the *professional* project manager from the masses of people with project manager titles. Many individuals with the project manager title are untrained and unskilled in professional project management methodology and tools.

The term "project manager" means different things to different people. It is difficult to tell the difference between project managers, program managers, and project coordinators. There is often little consistency in job descriptions, titles, rating methods, and pay across different project groups within a corporation as well as between companies. Job descriptions and related terminology are not standardized.

At the other end of the bell curve, a survey of benchmarking organizations found that about 20% indicated that they don't even have a project manager job title. One company uses the title "project administrator." Another uses engineering titles.

11.3.2. Pay Doesn't Reflect Responsibility

Often the project manager position is arbitrarily assigned a low management status in an organization. Companies have narrowly defined pay caps that limit the amount project managers can be paid. Benchmarkers describe situations where project managers are leading $100 million projects but receive less salary than vice-presidents responsible for managing a small operation with one or two employees. The pay for project managers in many large organizations is roughly the same for a $50,000 project as it is for a $50 million project. As project managers are assigned to larger and larger projects, frequently their pay does not correspondingly increase. Consequently, competitors who do recognize the value of the professional project manager can selectively hire the best project managers from large functional enterprises.

11.3.3. Project Manager Skills Don't Match Predefined Jobs

When considering large cross-functional projects, the job requirements and skill sets of the project manager are much broader than those found in most functional job descriptions. In Figure 11.2, it can be seen that as a typical manager progresses through life, their skill blocks become larger. When they enter the work force, they have little if any supervisory experience, advanced training, or leadership skills. At the pinnacle of their career, they have acquired advanced degrees, accumulated managerial experience, and developed their leadership skills. All of these skill blocks are independent of the particular job they are performing.

At the same time, the typical functional organization is approaching their job definitions in a somewhat parallel manner. Each job slot is defined according to the skills needed to perform it. After definition, all the jobs in the company are arbitrarily stuffed into the organization's job slots. (See Figure 11.3.)

In theory, the individual's skill blocks fit into the respective organizational job slots. (See Figure 11.4.) The problem arises when a person is arbitrarily placed in a job or when a job has not been defined.

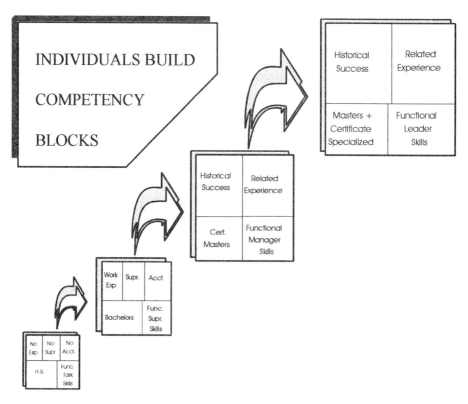

FIGURE 11.2

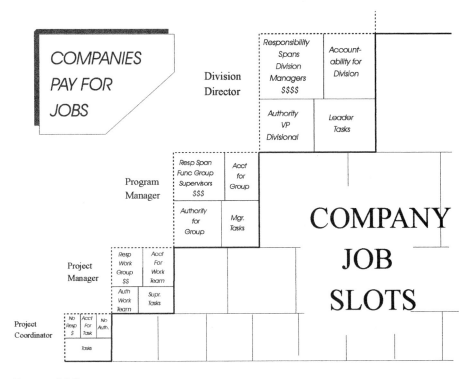

FIGURE 11.3

For example, the cross-functional nature of the professional project manager's role does not neatly fit with functional organization structures and associated job descriptions. As a result, professional project managers are sometimes arbitrarily inserted into a generic job classification.

Even when there is a good fit initially, as the project manager's training and experience increase, their skill block expands and they no longer fit the job slot. In many organizations, there are no corresponding job slots or pay grades for the advanced skill levels of the professional project managers.

11.3.4. Entrepreneurial Characteristics Create Broad Demand for Project Managers

The body of research evidence suggests that project managers share the same basic competencies as chief executive officers. The experienced project manager can just as easily run a business unit, profit

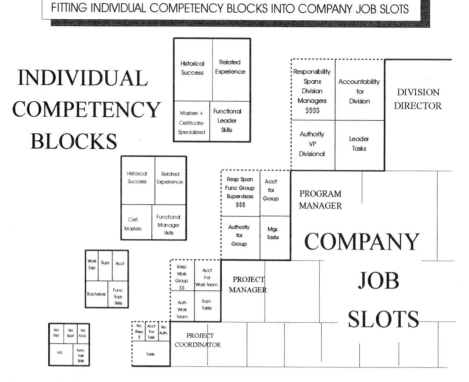

FIGURE 11.4

center, company, or any cross-functional organization. Several organizations in the benchmarking group have acknowledged the entrepreneurial nature of superior project managers and are using the role as a training ground for future executives. IBM, for example, requires project management certification before an individual can be promoted above a certain level in the company.

> **Standard:** **The best practices project organization recognizes that the project manager is effectively the chief executive officer (leader) of the project and exhibits the same actions, traits, and skills.**

The issue becomes important to the project organization as other functional areas and organizations become aware of project managers as an entrepreneurial resource. The consequence is that project managers can switch from managing projects to other types of management in the executive management track. In numerous examples related by benchmarkers, project managers were targeted as a valuable commodity and aggressively recruited by other corporate areas and companies. Companies report that their best project managers are hired away to manage business units and profit centers as well as by competition. The most competent and skilled project managers often have more financial incentives to move than to stay within the project management group and within the project manager career track.

This dilemma is a two edged sword. It is clearly beneficial for individual project managers and the profession in general to have two career paths available, i.e., the execution of larger and larger projects or sliding over to the senior management track. However, for organizations, it makes the issue of project manager pay, benefits, and other methods of motivating project managers more crucial.

11.3.5. Pay Doesn't Keep Pace with Organization Demand for Higher Professionalism

As industry recognizes the benefits of project management in improving the bottom line, increasingly broadened is the scope, size, and complexity of projects. In response, the standards of project manager professionalism are raised. The project group invests time, effort, and money to train and nurture professional project managers. Sometimes pay increases at a slower pace. Again, the project manager is then targeted as a valuable asset or commodity by other corporate areas and companies who hire them away.

11.3.6. Project Manager Turnover

Some benchmarking organizations have responded to the increasing salaries for experienced project managers by hiring lower paid college graduates. They recognize the new graduates have limited experience. Consequently, the new project manager is extensively trained in project management methodology, encouraged to become professionally certified, and provided an experience base consisting of internal projects. The problem arises because at the conclusion of this process, the professional project manager has become a more valuable asset than when they were employed. Consequently, they become inviting targets for competitors to hire away.

11.3.7. Problematic Performance Evaluations

The project manager's performance is often measured on the basis of the *project's success,* and success criteria are loosely defined and vague. In many cases, the project manager has little influence over which projects they are assigned. In addition, new project managers do not have a history of prior performance on which to evaluate their success. In these cases, it is necessary to evaluate performance based on how well the project manager runs the project, not the project's success. There is a need to define how to evaluate the performance of project managers, set objectives for project managers, and measure the quality of project manager performance.

11.3.8. Lack of Precedent for Project
Manager Compensation

Benchmarkers report that human resource departments are reluctant to give project managers bonuses or competency or team based performance pay because they don't see similar actions in other companies. Corporate culture is a complicating factor. The question is asked, "Why should one group be treated differently?" "If there is a need, how can it be equitably accomplished?"

References: Toney and Powers, 1998.

11.4. SOLUTIONS

If there were a room full of project managers they would likely say, "The solutions to the compensation and motivation problems are easy; pay us more money and give us a clear career path to the top!" In response, best practice project organizations and their host organizations generally *agree* with the benchmarkers' suggestions. In fact, best practice project organizations report that they *are* paying more money, developing job descriptions and multiple career tracks, and working on ways to more accurately evaluate project managers' performance. A discussion of the approaches taken by benchmarking organizations to each of these topics is discussed.

11.4.1. Organizational Approaches to Pay

Best practices project organizations are constantly searching for the most effective ways of rewarding and motivating project managers. They benchmark with other organizations to determine ways the existing pay system can be improved and how the pay structure contributes to project goal achievement, project manager motivation, and improved team effort toward achieving the project's goals.

Benchmarking organizations surveyed utilize two basic approaches to pay, straight salaries and bonuses based on performance.

11.4.1.1. Traditional Salaries

The salary category includes the traditional job based pay where a fixed periodic amount is paid for performing a defined job. The survey of benchmarking organizations determined that about one-third use salaries based on tiered pay categories. Project managers are limited in their maximum pay by the category and the width of the pay range.

In some cases, benchmarkers assert that straight pay is the most appropriate form of remuneration. One example "bail out," "smoke jumper," and other high risk or uncertain projects that are subject to the impact of numerous uncontrollable factors. Another is where project managers have the primary objective of simply trying to reduce losses. In situations such as these, benchmarkers say that a reasonable solution is to provide the project manager sufficient pay to compensate for the risk involved.

Several organizations improve the flexibility of salary based pay by reducing the number of pay categories. The range of pay available for the remaining categories is expanded, or "broad banded." Thus, project managers can rise to higher levels of salary without being reclassified. They can remain on a project manager track and be rewarded for performing larger and larger projects.

In theory, broad banding provides greater flexibility to reward project managers for increased responsibility, experience and expertise. Anecdotal experience by participants hints that with broad banding, most of the salaries are clustered around the middle of each pay category with the result that the concept makes little difference in most individual's pay.

11.4.1.2. Bonus Based Pay

Organizations report favorable results from providing bonuses to motivate and reward project managers and teams. The bonuses are a reflection of (a) the competency or skill applied by the project manager and/or (b) preagreed upon goals or team performance targets. A survey conducted by Lawler, Ledgord, and Change found that about one quarter of companies responding indicated they have implemented efforts to make project manager pay more reflective of competency or performance. Of the respondents to their survey, 60% said it was very successful in increasing performance. Associated benefits were improvements in employee flexibility, more easily facilitated job rotation, and the opportunity for pay improvement when promotions were not available. Approximately 53% said their organizations would use performance and competency based pay more in the future.

General Mills reported that the higher average pay for individual workers was more than compensated for by bottom line benefits of production increases, leaner staffing, and fewer accidents. Honeywell judged that performance and competency based pay made them more competitive in the high competition labor market, improved safety, and increased productivity and provided greater employment stability.

Proponents of paying bonuses, rewarding performance and the superior application of competencies, list numerous benefits in comparison to the practice of paying straight salaries. The rewards more closely mirror bottom line benefits to the company. Investment in project manager training has a more observable payback for the company *and* the individual. The organization is more likely to retain the top performers. Bonuses and performance based pay serve as an enticement to attract top performers. Clearly differentiated are the superior project managers compared to less skilled individuals that hold the project manager title. The consistently most expensive (i.e., most competent and highest performing) project managers could be assigned to the largest and most complex jobs. Hence, rewards would more closely match responsibilities.

- **Competency, best practices–or skill-based pay.** About 10% of benchmark organizations surveyed rely upon a competency-, best practices–, or skill-based pay structure. These pay systems reward the individual for application of competencies and best practices as project managers. They tend to separate the project manager's performance from whether the project was successful or not. They reflect the best practices applied to the particular project. They also reward skill development as well as paying for the job performed.

  The approach is particularly appropriate when the project manager is arbitrarily assigned a project or when there are numerous external influences beyond the control of the project manager. In many organizations the project manager has little to say about the project they are assigned to lead. Some projects are marginal and face failure from the day they start. In other cases project managers are assigned to projects that are already failing. In all these situations, it would be unfair to evaluate performance and pay solely on the basis of the successful conclusion of the project. Competency based pay involves a periodic evaluation of the quality of managing the project (even if the project itself is failing) and payment of bonuses as a reflection of the performance.

- **Team based performance pay.** In an era where there is emphasis on the team approach, it is axiomatic that most pay systems continue to primarily reward individual performance. In response, several companies are working to develop project team based pay. As an incentive for project managers, bonuses based on team performance are used by about 18% of benchmark survey respondents. Results are generally favorable. Electronic Data Systems reports that the bonus program is successful to the point that the project group is attracting project managers from other corporate project groups. Results of group performance pay plans include improved productivity, better teamwork, increased pay satisfaction, enhanced communications, and reduced numbers of grievances.

Group pay is a two part process; setting goals and then rewarding goal achievement. Goal attainment performance metrics are agreed upon between the project manager/team and the project organization. The team receives its bonus when the metrics are attained. The metrics or thresholds establish the minimum acceptable performance necessary for success. Goals that are specific, observable, and measurable raise performance higher than "do-your-best" type less specific goals.

Benchmarkers say that goals must be normally attainable, meaning they should be established at *modest* levels. In some cases the team is rewarded progressively as a reflection of performance gains. For example, the reward may be based on how quickly the project is completed. The earlier the completion date, the higher the bonus.

Bonuses represent a form of feedback to the team. Bonuses should be sizeable, paid frequently, and closely associated with the pertinent actions to provide a greater incentive intensity and result in improved performance levels. Research also has determined that groups with frequently paid bonuses tend to set more difficult and challenging goals and are more committed to the goals once they are set. Size of bonus increases satisfaction, commitment, motivation, and performance.

There are numerous varieties of bonus plans. They range from conventional bonuses paid for good performance to more complex bonus arrangements. One program staggers bonuses over a three-year period. The objective is to provide incentives for employees to remain with the company. Another benchmarking organization reports that 27% of each individual's pay is team based. Meeting objectives such as bringing the project in on time and budget serve as measures of success.

One organization in the benchmarking group includes components of (a) base salary, (b) at-risk pay, resulting from employee and manager established expectations, goals, and self-improvement (up to 20% of pay), and (c) a project related bonus. The specific amount of the bonus is determined by the project organization in combination with the project manager and team.

A third program is a combination of at-risk pay as determined by (a) overall corporate performance, (b) service performance with the group's geographical area, and, (c) achievement of corporate strategic goals.

Benchmarkers warn that a common problem reported is that even when people fail to achieve goals for reasons that were under their control, they often still want the bonus. For goal based pay to be effective, attendees conclude that its administration is a delicate process.

References: Hollensbe and Guthrie, 2000; DeMatteo and Sundstrom, 1998; Wageman and Baker, 1997; Lawler and Chang, 1993; Schuster and Zingheim, 1992.

11.4.2. Promote Pay Programs to Human Resources Departments

Benchmarkers caution that in most organizations it is not a simple matter for the project organization itself to adopt new approaches to pay. Usually pay administration is the role of the human resources department. Consequently, participants voice the need to present the human resources department with research literature and examples from other best practices organizations in an effort to communicate the bottom line benefits and motivate the desired changes.

11.4.3. Project Organization Job Descriptions

The need for clear job descriptions and statements of career progression was summed up by one participant, "It should be made clear to everyone how they can progress, what skills need to be developed, and how the skills can be acquired." To accomplish this, it is stressed that there must be descriptions and a career progression guide. For example, at one company, project managers have four clearly defined job levels. Another organization uses five distinct levels of project management job descriptions. The job descriptions are used in conjunction with a career enhancement program that encourages development and retention of project managers.

> **Standard:** **The best practices project organizations develop clear and precise job descriptions to differentiate different types of project manager roles.**

11.4.3.1. Typical Project Group Organization Chart

Figure 11.5 depicts project management jobs and their position within the organization. Ideally, the group reports to a multifunctional, senior level executive or committee. In some cases where large portfolios are managed, the project director reports directly to the chief executive officer of the company. The project organization and entire project portfolio is managed and administered by a director level individual. The project director has responsibility for multiple programs, program managers, project managers, and administrative

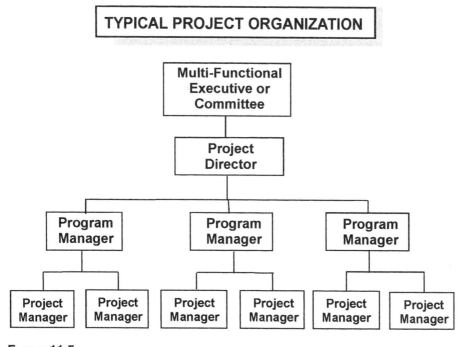

FIGURE 11.5

activities. Reporting to the project director, are the program managers who supervise multiple projects and project managers. The individual projects are under the leadership of a project manager. The projects are composed of cross-functional teams.

Not shown in the figure is the Project Coordinator position. This individual reports to a functional manager and is associated with matrix-structured projects. The project coordinator typically facilitates smaller projects involving one or two functional areas. Responsibility for project success lies with the functional manager, not the project coordinator. Conflict resolution, leadership, and coaching is provided by functional management.

11.4.3.2. Project Manager Job Matrixes

Table 11.1 defines typical skill sets, pay range, scope of responsibility, and performance deliverables for four skill blocks corresponding to project organization positions.

11.4.4. Programs to Retain Superior Project Managers

There is general agreement that project manager retention is an issue. Some participants have formal programs to retain project managers. The programs consist of measuring prior turnover and developing strategies to reduce future turnover. Solutions generally include providing multiple career track opportunities and more flexible pay programs. Also included are activities that build a sense of corporate culture and family. A broad variety of benefits can serve as important motivational tools. Project organizations describe perquisites such as stock options, a productive and pleasant work environment, educational opportunities, and individualized rewards and incentives. One benchmarker said, "At my company we ask employees what *they* want in the way of rewards; i.e., travel, bonus days, education. Then we try to tailor benefits to the individual's needs." Another benchmarker adds that excellent benefits and "having fun" as part of the job, contribute greatly to employee satisfaction.

> **Standard: The best practices project organization has a program aimed specifically at retaining people.**

Retention programs in other industries have been successful to the point that turnover has been reduced by half. Participants emphasize that the retention program should be a continuing process. Often, short-term or "one-shot" approaches are viewed with cynicism by employees who see them as transparent efforts to buy loyalty.

11.4.5. Ambassadorial Role of Project Organization

A clear pattern of successful project organizations is supporting the project teams through emphasis on ambassadorial, impression management, public relations, promotional, and political activities and conflict resolution. Ambassador activities include lobbying for support and resources and buffering the project teams from outside pressures. Ambassadorial activities enhance project visibility and outsiders' perceptions of team competency.

The value of the ambassadorial role is supported by the literature. Dutton and Ashford found that issue selling has a positive correlation with organizational performance. Promotional activities enhance visibility and outsiders' perceptions of competency and power. Effectively the project organization attempts to control external images of itself and the projects being executed.

For ambassadorial activities to remain effective, they must be followed up with performance. Ancona and Caldwell determined that an *excess* of ambassadorial activities is associated with *reduced* performance. Ambassadorial activities achieve positive outcomes in the short term. Long term, the most successful teams manage both the external public relations and the project workflow for superior results.

Standard: **The best practices project organization serves an effective ambassadorial role that represents, protects, and advances the interests of the project teams with outside political groups.**

The research of Dutton and Ashford showed a positive correlation between successful "issue selling" and team performance. Ancona and Caldwell found that promotional activities enhance visibility and outsiders' perceptions of competency and power. The more effective groups attempted to control the external image of themselves through impression management.

TABLE 11.1 Typical Project Organization Position Descriptions

	Skill block 4: Project Director	Skill block 3: Program Manager	Skill block 2: Project Manager	Skill block 1: Project Coordinator
Amount and degree of historical success rate vs. risk	High historical success related to risk as a program manager	High historical success related to risk as a project manager	High historical success related to risk as a project manager	High historical success related to risk as a project coordinator
Technical qualifications Education	Master's or Doctorate Masters Degree	Masters Degree	Bachelor's Degree	Bachelor's in progress
Specialized education	Post graduate level, leadership, project management, entrepreneurship, courses	Master's level project management	Project management courses in progress	None
Certification	PMP, IMC, or equivalent	PMP	PMP in progress	None
Experience required	1 year project director 1 year program manager 1 year project manager	1 year program manager 1 year project manager	1 year program manager 1 year project team member	None
Application tools and skill expertise required	Leadership skills, communication tools, presentation skills, financial analysis	Project management software overview, leadership skills, communication skills, financial analysis	Project management software in depth understanding, leadership skills, communication skills	None
Pay range, industry	$85,000–$250,000	$65,000–$175,000	$25,000–$110,000	$18,000–$40,000
Size of project or program	$1.0–2.5 billion	$100–$500 million	$50,000–$100 million	$5,000–$10 million
Span of responsibility				
Number of functional areas	All	All	All	One to three
Number of employees	4 to 20 Program Managers	4 to 20 Project Managers	Four to hundreds of team members	4–100 associates

	Vice-President or Senior Executive	Program Director	Program Manager	Functional Manager
Reports to	Executive			
Primary stakeholder interfaces	Senior Executives Program Managers, major clients	Project Director, Project Managers, clients	Program Manager, clients, project team members	Functional Manager, project associates
Decision accountability	Complete project group performance accountability	Program performance accountability	Project performance accountability	Accountable for coordinating activities
Formal authority	Upper middle manager to lower level vice-president	Upper middle manager	Middle to upper middle manager	Coordinator
Performance deliverables	Project group goal achievement; Strategic planning; Personnel planning and recruitment; Project initiation; Initial client contact; Project cost benefit analysis; Accounting and profitability; General project management training; Establish and teach methodology; Project auditing; Supervise program managers; Evaluate project manager performance; Group conflict resolution	Program goal achievement; Supervise project managers; Program planning, control and leadership; Project manager performance evaluation; Program priority establishment; Program conflict resolution; Communication with program clients; Support methodology, communications, and training; Implement group strategic plan; Project cost, benefit, and control tracking	Project goal achievement; Apply standard approaches and procedures; Meet time, cost, quality, and scope goals; Provide project planning, control, and leadership; Coach and consult with team members; Communicate project specifics with stakeholders; Resolve project conflicts	Coordinate project activities; Meet time, quality, and scope goals; Provide project plans and control; Communicate with functional leadership

The extensive research of Ancona and Caldwell quantified the positive impact of specific ambassadorial activities on project goal achievement. They found a high correlation in the +.7 range for the activities of "absorbing outside pressures for the team so it can work free of interference," "protecting the team from outside interference," and "preventing outsiders from overloading the team with too much information or too many requests." Activities recorded in the +.6 range included , "persuading other individuals that the team's activities are important," "scanning the environment inside the organization for threats," "negotiating with others for delivery deadlines," and "Talking up the team to outsiders." In the lower but still significant range of +.5 were "persuading others to support the team," "acquiring resources (e.g., money, members, equipment) for the team," "finding out whether others in the company support or oppose the team's activities," and "finding out information on the company's strategy or political situation that may affect the project."

11.4.5.1. Political Skills

Benchmark forum participants continually stress that a key element impacting project speed and success is the ability of the project organization to successfully interface with an entrenched host organizational culture, bureaucracy, and political system. The burden to achieve success in this arena rests largely with the project organization. Project organization managers should be good politicians and ambassadors. They should buffer the project teams from outside influences. They should engage in presenting a positive but realistic picture to stakeholders and ensure resources are made available.

Failure to understand the political aspects of the project can lead to project speed and efficiency reductions and even outright project termination. Ingram's data showed that there is sometimes a tendency for project leaders to think of projects as independent groups whose success depends mainly on the technical quality of the work performed and the professionalism of tools used such as quantitative and rational schedules, nicely prepared Gantt charts, resource leveling, cost-benefit analysis, and completed templates. The reality is that project success was often dependent upon other more complex human and group factors. Projects are political organizations within larger and multiple other political organizations. Ingram found that project leaders, team members, stakeholders, and the myriad of political groups were often primarily motivated by their own self-interests which were in direct conflict with the objectives of the project.

11.4.5.2. Team Buffering

The superior project organization serves as a buffer or "gatekeeper" between the project teams and the external political environment. Brown and Eisenhardt found that the gatekeeper function is an important facet of project success. The superior project organization impacts project performance by increasing the amount, variety, and diversity of information available to the project teams. The gatekeeper function gathers, understands, and translates information into more meaningful and useful information for team members. As a consequence, gatekeepers are extremely valuable in increasing the probability of project goal achievement, suggest researchers Allen and Cohen.

11.4.5.3. Lobbying for the Project Teams

The ability of the project manager to communicate the needs, interests, and accomplishments of the project teams to outsiders and stakeholders is an element of project organization success. A successful project requires good people, facilities, tools, and other resources said Gupta and Wilemon. Lobbying attains these needs.

> **Standard:** **The best practices project organization lobbies for resources and project needs.**

11.4.5.4. Communicating Project Benefits

The success of project management groups in large functional organizations is not simply the result of open-arm acceptance by peer groups and superiors. It involves an effort to communicate and promote the benefits of the project group and individual projects. This is not necessarily a self-serving approach, but is designed to inform others of the ways that the project group and teams have improved the ability of the organization to achieve goals. Project speed and efficiency is maintained by reducing the need to constantly justify activities and search for resources.

11.4.5.5. Managing Conflict

The project organization encourages internal project related conflicts to be resolved by the project manager. Occasionally conflicts arise that involve individual projects and outside parties, stakeholders, or other project teams. Conflicts can have a negative psychological as well as monetary impact on the projects. Most benchmarking groups have no formal process to address conflicts. About 20% do.

Two generalized conflict related factors have direct impact on achieving project success, judged the research team of Hayes. First is the need to bring conflicts to the surface early, and second is the need to resolve conflicts through mutual accommodation at low levels in the hierarchy. In either case unresolved conflicts can disrupt project speed, efficiency, and effectiveness.

Standard: **The best practices project organization resolves conflicts early and at the lowest possible levels.**

Most often, conflict resolution within the team rests upon the shoulders of the project manager. The majority of conflicts are handled in an informal fashion. For more serious conflicts, some best practices organizations have formal conflict management procedures. For example, some of the benchmarking organizations report success with a "48 hour conflict resolution commitment." Antagonists attempt to resolve the conflict at the lowest level. If unsuccessful, the conflict is taken to the next higher level of management. If it is not resolved within 48 hours, the conflict proceeds to the next level. According to respondents, it is rare that problems ever ascend over one or two levels. It is noted that a danger of the approach is that antagonists sometimes bury a problem rather than face higher levels of management. In those cases the root problem causes remain unresolved.

The benchmarking forum attendees and researchers agree that there is one area of conflict that demands constant monitoring and an immediate response by the superior project organization. The relationship and association of the project organization or specific project teams with the host organization and its imbedded culture represent a major area of potential conflict that can have a negative impact on project goal achievement.

Project managers have a primary objective of meeting time and budget targets and achieving project objectives. Invariably they are working in a environment heavily influenced by functional managers who have ongoing activities to perform that are unrelated to the project but just as important and time consuming. By its nature the project intrudes upon the functional areas. Projects often appear to "commandeer" personnel and other resources and interfere with functional area routines. If the functional area is a revenue or production producer, the project group activities could have negative impact on sales efforts and production output. Further, the successful completion of the project tends to result in accolades for the project leadership and team. In the meantime, the functional areas are performing their tasks with seemingly less acknowledgement. All of these differing elements and objectives create a natural conflict between project and functional management.

Kingdon concluded that any organization that balances multifunctional projects with the interests of functional departments would encounter three fundamental areas of conflict. The first involves work and potential solution related technical decisions. Often the work of the project team distracts from the day-to-day activities of the functional departments. Second are results related to team member salaries and career management. To have highest effectiveness in managing and motivating team members, the project manager would ideally have full authority over pay and remuneration. In addition, companies are increasing project manager pay and recognizing successful project leadership as displaying potential for executive leadership, often in preference to traditional functional department approaches. Finally, decisions are related to staffing and assignment of team members to specific project activities. Personnel for project teams typically come from the functional organizations. Often the project has high visibility, referent authority, and status from senior management and is working on interesting activities that directly relate to organizational strategy. As a result there is a tendency for project managers to have leverage in selecting people from the functional areas. The end result is conflict.

The potential impact of the natural tendency to have conflict between project teams and functional organizations cannot be overstated. Benchmarking forum participants stress that conflicts with the host organization can and have resulted in the termination of project teams and entire project organizations. Conflicts with the host organization are particularly delicate because they are based on strong and entrenched cultural differences and perceived threats to the careers and well-being of people who are often at high levels of the organization. Benchmark participants have repeatedly reported that the more successful the project organization, the more it is perceived as a threat to many members of the functional organization.

The role of the project organization is to diplomatically balance the interests of both functional areas and the project teams. Consequently one of the responsibilities of the project manager becomes that of the ambassador to ensure effective operations.

References: Ingram, 1998; Brown and Eisenhardt, 1995; Ancona and Caldwell, 1992b; Thomas, 1992; Clark and Fujimoto, 1991; Ancona and Caldwell, 1990; Gupta and Wilemon, 1990; Hayes et al., 1988; Allen, 1977; Kingdon, 1973; Cleland, 1999.

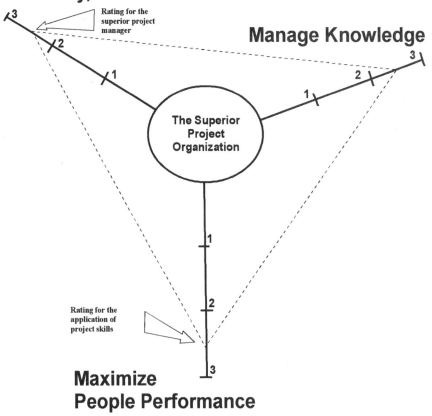

PROJECT ORGANIZATION COMPETENCY SUB-WHEEL
Support of the Project Goal Achievement

Structure for Speed, Efficiency, Effectiveness

Rating for the superior project manager

Manage Knowledge

The Superior Project Organization

Rating for the application of project skills

Maximize People Performance

FIGURE 11.6

11.5   MEASUREMENT OF PROJECT ORGANIZATION SUPPORT OF PROJECT GOAL ACHIEVEMENT

An evaluation of a project organization's level of support for overall project goal achievement can be numerically quantified using the Organizational Support Competencies Subwheel. The numerical results from this wheel are rolled up or transferred to the Master Project Organization Competency Wheel. In Figure 11.6, it can be seen that the project organization's support of project goal achievement is comprised of three elements: structuring the project teams for speed, efficiency, and effectiveness; managing knowledge through methodologies, tools, and training; and maximizing people performance.

Ranking Scale

Frequently, if not always	Fairly often	Sometimes	Once in awhile	Not at all
5	4	3	2	1

EVALUATION QUESTIONS

Chapter 8: Managing For Goal Achieving Speed, Efficiency, and Effectiveness

5 4 3 2 1   1. The project organization emphasizes project speed in achieving goals, while maintaining efficiency and effectiveness.

5 4 3 2 1   2. The project organization structures the project teams for speed, efficiency, and effectiveness.

5 4 3 2 1   3. The project organization organizes projects into small groups, subgroups, and modules.

Chapter 9: Knowledge Management: Providing Structured and Predictable Methodologies and Tools

5 4 3 2 1   4. The project organization develops and applies a predictable methodology to manage projects.

5 4 3 2 1   5. The project organization uses a methodology that is broad and flexible.

5 4 3 2 1   6. The project management methodology emphasizes project and people management over tools during the execution and control phase of the project.

5 4 3 2 1   7. The project methodology includes a project charter that contains all material information needed to make a decision whether to approve the project.

5 4 3 2 1   8. The project organization encourages an *appropriate* level of detail in project plans.

5 4 3 2 1   9. The project organization requires definition of project scope including all aspects that involve additional resources if changed.

5 4 3 2 1   10. The project methodology requires a graphical work breakdown structure for the project that shows an appropriate detail of deliverables and activities.

5 4 3 2 1   11. The project methodology includes a schedule showing all appropriate activities, sequences, durations and the critical path.

5 4 3 2 1   12. The project organization encourages the use of numerous milestones to serve as goals and against which to measure schedule performance.

5 4 3 2 1   13. The project organization uses contingency funds and queuing buffers to ensure schedule and budget accuracy and to factor in risk and uncertainties.

5 4 3 2 1   14. The project methodology includes a project budget (or cash flow) that is appropriately accurate for the scope of the project.

5 4 3 2 1   15. The project methodology identifies and analyzes project specific risk events, determines their likelihood of occurrence and significance, and plans appropriate responses.

5 4 3 2 1   16. The project methodology prepares an appropriately detailed set of functional and technical specifications for the project product or service.

5 4 3 2 1   17. The project methodology prepares a communications plan that provides appropriate frequency and detail for the needs of each stakeholder group.

5 4 3 2 1   18. The project methodology identifies customer requirements and measures project performance against those requirements.

5 4 3 2 1   19. The project methodology includes a kickoff meeting.

5 4 3 2 1   20. The project methodology includes the use of variance analysis as a tool to control deviations from the project plan.

5 4 3 2 1   21. The project methodology manages project scope changes as a formal process.

5 4 3 2 1   22. The project methodology emphasizes formal presentation of the final deliverable to the customer.

5 4 3 2 1   23. The project methodology includes a termination checklist to ensure all termination activities are performed.

5 4 3 2 1   24. The project organization contributes to the organization's project knowledge base by maintaining appropriate project documentation (lessons learned and journals and the charter and plans) in a library or depository.

5 4 3 2 1   25. The project methodology includes measurement and communication of project contributions to stakeholders and the organization.

Chapter 10: Knowledge Management: Providing Tools Standards and Training

5 4 3 2 1   26. The project organization provides state-of-the-art software, information, communication, and project tools.

5 4 3 2 1   27. The project organization has a personal development and training program based on identification of skills and competencies needed by the individual or group.

5 4 3 2 1   28. The project organization measures and quantifies the value of training..

5 4 3 2 1   29. The project organization commits to lifelong learning for project managers and teams.

5 4 3 2 1   30. The project organization builds excellence and high quality into every activity performed.

5 4 3 2 1   31. The project organization benchmarks and implements best practices from other individuals, teams, and organizations.

5 4 3 2 1   32. The project organization acts as a role model to communicate positive values and guide the behavior and performance of the team.

5 4 3 2 1   33. The project organizations take a proactive stance in developing programs aimed at attracting, selecting, retaining, and rewarding top performing project managers.

Chapter 11: Supporting Project Manager Performance

5 4 3 2 1   34. The project organization provides supportive, facilitative and servant leadership for its project managers and teams.

5 4 3 2 1   35. The project organization hires superior project managers and those with potential to be superior project managers.

5 4 3 2 1   36. The project organization hires project managers who possess a minimum of a four-year college degree or equivalency.

5 4 3 2 1   37. The project organization hires project managers who possess the PMP (Project Management Professional) certification.

5 4 3 2 1   38. The project organization values lessons learned from prior failures in project manager candidates.

5 4 3 2 1   39. The project organization recognizes that the project manager is effectively, the chief executive officer (leader) of the project and exhibits the same actions, traits, and skills.

5 4 3 2 1   40. The project organization develops clear and precise job descriptions to differentiate different types of project manager roles.

5 4 3 2 1   41. The project organization has a program aimed specifically at retaining people.

5 4 3 2 1   42. The project organization serves an ambassadorial role that represents, protects, and advances the interests of the project teams with outside political groups.

5 4 3 2 1   43. The project manager buffers the team from external pressures, demands, and politics

5 4 3 2 1   44. The project organization lobbies for resources and project needs.

5 4 3 2 1   45. The project organization resolves conflicts early and at the lowest possible levels.

To determine the number of points to apply to the Master Organizational Competency Wheel, score the questions as follows:

	Maximum Rating
Questions 1–3: multiply the points selected by 1.1	3.3
Questions 4–33: add the points and multiply by 0.118	3.3
Questions 27—45: add the points and multiply by 0.194	3.3

Appendix A

Benchmarking Forum Background and Supporting Information

A.1. THE TOP 500 PROJECT MANAGEMENT BENCHMARKING FORUM

This work would not be possible without the involvement and support of the Top 500 Project Management Benchmarking Forum. The forum is administered by the nonprofit Executive Initiative Institute and composed of project management executives from approximately 60 large for-profit and governmental organizations. The objectives of the group are twofold: first to *identify* best practices associated with superior project managers and project groups in large functional organizations and, second, to encourage participants and industry in general to *implement* the best practices and attain the associated benefits.

Benchmark forum meetings are held approximately every three to six months. The forums are designed to encourage participants to interact freely; find answers for immediate problems; share new ideas, concepts and concerns; and return to their respective organizations with a wealth of information that can be immediately applied to improve project performance.

The forums follow a scientific benchmarking approach. Each participant is asked to list problems relating to the implementation of project management in their organization that they wish to benchmark with their peers from other organizations. From this, a list of generalized subject areas is developed. Roundtable discussions (a) define, compare, and contrast how each participant's organization addresses the specific problems, (b) identify and agree upon key success factors and best practices to resolve the problems, and then (c) discuss ways of applying the findings to one's own workplace. Participants also present ways their specific organizations excel at project management related processes and functions as well as bring examples of new and effective project management tools. The conclusions reached in the forum are treated as hypotheses and are supported with literature searches, surveys, questionnaires, and in-depth interviews with members and their associates. At follow-up forums, participants report their degree of success in applying the best practices.

From all this, a list of generally agreed upon competencies, best practices, and standards; along with a description of the supporting research, is generated and distributed.

Numerous articles have been written about the research. In 1996 the group published a book entitled *Best Practices of Project Management Groups in Large Functional Organizations*. The book was well received by industry and the project management profession and helped guide many organizations to improved efficiency. This document builds upon that foundation and adds the details of the expanded and more intensive and detailed research results of the last three years.

A.2. DEFINITIONS

To establish standards and define best practices of superior project organizations, it is necessary to have a clear understanding of the terms that describe the project organization's attributes. There is confusion about terms such as "competencies," "best practices," and "standards." Listed here are definitions of the terms as used by the Top 500 Benchmarking Forum.

- **Competencies.** Competencies are defined as a *generalized* category of actions, traits, and skills applied by superior project organizations. For example, a key competency is honesty.
- **Best practices.** Best practices are the *specific*, observable and measurable actions taken that result in the achievement of project goals. They are statements or guidelines that apply

to a theoretical "best practice" or "superior" project organization. A best practice associated with the competency of honesty is, "The best practice project organization is truthful in all dealings and relationships." The term "best practice" is often used interchangeably with the term "competency."

- **Standards.** As used in this book, standards are the overarching rules, principles, practices, models, and measures established by the Top 500 Project Management Benchmarking Forum. They reflect the authority of the Top 500 Project Management Benchmarking Forum as a recognized industry standards-setting body. Standards are available, familiar and regularly and widely used in the project management field. They are substantially uniform and well established by usage in the speech and writing of the educated within the field. They establish benchmarks of performance.

A.3. CHARACTERISTICS OF COMPETENCIES

Competencies have several distinguishing characteristics. They tend to define what a person or organization should do rather than what they should know. They characterize the manner in which the individual or organization performs the job rather than being reflective of the success of the project or group. Competencies are observable and measurable and drive superior rather than average performance. They have been scientifically validated and are weighted in terms of their importance in increasing the probability of project goal achievement.

Questions are asked about the difference between the *Project Management Body of Knowledge Guide (PMBOK Guide)* and project manager standards and competencies. The difference is that the *PMBOK Guide* describes a body of knowledge about a subject area (e.g., project management) and competencies describe actions, practices, traits, and skills of individuals and groups. Simply stated, the PMBOK *Guide* details what a project manager *should know*; competencies define what they *should do*.

A.3.1. Impact on Goal Achievement

Some researchers have associated competencies with successful goal achievement. For the purpose of defining project manager competency, a broader definition is necessitated. It is possible for a person to perform the project manager's role in a highly competent fashion, yet not achieve the group's goal. Tying competence to goal achievement limits the definition to projects that have been completed and a determination made whether the goal was achieved.

In many organizational environments, there is a need to evaluate the leadership competence of people who have yet to attain goals, are in training programs, or are arbitrarily assigned to leadership roles in projects that are subject to outside forces and constraints. For example, often leaders are assigned or forced into projects that could be categorized as losers from the beginning. If the leader applies goal-achieving methodology in a skillful manner, their performance is not necessarily a reflection of the project's success.

For all these reasons, project manager competency is not tied to successful project goal achievement. In the preceding cases, and any other time the goal is in the future, it is necessary to evaluate the leader's application of best practices and overall competencies.

A.3.2. Observable and Measurable

Competencies are a collection of observable and measurable factors that require no inference, assumption, nor interpretation.

A.3.3. Drive Superior, Not Average, Performance

Competencies should consistently predict and distinguish superior from average performance.

A.3.4. Scientifically Validated

From an ethical, moral, scientific, and legal view, competencies *must* be validated. They are used for hiring, promotion, training, performance appraisal, and testing—all activities that have major impact on individuals and organizations. The validation process is as follows.

A.3.4.1. Hypothesis

Competency validation is an application of the scientific process. It assumes that there is a theoretical "one best way" to attain objectives. Its hypothesis is that, *there is a methodology for leadership success consisting of measurable competencies, that, when applied in varying degrees of intensity, increase the probability of organizational goal achievement.*

A.3.4.2. Competency Identification

It is necessary to find and identify specific competencies to evaluate. Conducting exploratory research, interviewing superior project managers and experts in the field, and executing a literature search and in-depth review of prior research accomplishes the task.

A.3.4.2.1. Exploratory Panel of Participants. Some organizations use a panel of participants or "experts" in the exploratory stage of their competency investigation. The panel members list the competencies that they think are important. Examples of the panel of participants of project manager professionals are the Top 500 Project Management Benchmarking Forum or a committee of managers from within the company.

The exploratory panel is the first step in the competency identification process. It must be followed by the remaining steps of the validation process for the competencies to be considered scientifically valid. Benchmarkers report that a tendency is for organizations to conduct the competency identification process in a brainstorming or committee atmosphere and then to consider their efforts complete. There are dangers associated with such an approach.

Many organizations develop a list of competencies from a group process and then proceed to the development of a questionnaire or other evaluation tool. As a result, the competencies are little more than the opinions of a group of executives or managers. Usually, the perceived competencies are not weighted. Project manager or organizational performance in nonmaterial areas is given equal weight with material areas. In other cases, important competencies are overlooked or ignored.

Another criticism of the approach is that the committee tends to establish competencies and associated tests that replicate the characteristics of the committee members. Project managers who mirror the characteristics of the selection committee are then promoted. The subsequent perception of success of the process becomes a selffulfilling prophecy. In addition to the moral implications of negatively impacting people's lives, the unvalidated competencies open the companies to criticism and possible legal action. Development of *validated* competencies provides companies and researchers a solid research based foundation upon which to base their evaluations.

A.3.4.2.2. Exploratory Interviews of Project Organization Leaders. Exploratory interviews attempt to determine the views of successful project organization managers. This method has the advantage that the results of several interviews can be quantified and various factors can be ranked. The major problem with the approach is that often leaders describe only actions taken and rarely address their personal characteristics such as truthfulness. In other words, the leader querying process determines what they think they do and may not accurately reflect what they actually do. To determine this, a literature search and questionnaire are required.

A.3.4.2.3. Literature Search. The literature search gives emphasis to scientific studies that develop empirical relationships between specific competencies and goal achievement.

A.3.4.3. Questionnaires

The information gathered from the literature search, panel discussions, and interviews of project managers and project manager experts is compiled into a series of test questions. The questionnaire replicates, expands, and builds upon prior research and information gleaned from the panels of participants and leaders. Its ultimate objective is (a) to obtain quantifiable results and rankings that describe what the project manager does to achieve goals and the differences between superior project managers and everyone else and (b) to provide a numerical weighting of each factor in terms of its impact on goal achievement.

A.3.4.4. Workplace Testing

Finally, the findings of the research are tested in the workplace and validated by the test of time. This historical workplace validation and experience can be initially accumulated with workplace testing. However, general leadership and managerial practices may require years of usage before they become generally accepted.

A.3.5. Weighted for Impact on Project Goal Achievement

Competencies have varying degrees of influence on project goal achievement, hence they should have different weights or values. For example, the practice of being absolutely truthful in all dealings and relationships has one of the highest correlations with project goal achievement. Many consider honesty to be a mandatory characteristic of superior project managers. On the other hand, competencies such as using computerized scheduling tools and software have been shown to improve project goal achievement but are not mandatory. Projects were conducted quite successfully for millennia before personal computers and sophisticated project scheduling software were invented.

A.4. IMPORTANT COMPETENCY CONSIDERATIONS

A.4.1. Might Not Pertain to Every Situation

In discussions about the results of the research with project managers, the most common response is, "Your findings don't apply to me or my organization. We are in a different situation or use a different organizational approach." From the standpoint of statistical accuracy, their responses are correct. The study and other findings in this book deal with averages—each reader or their organization is an individual. Consequently, this book does not specifically pertain to each individual reader or organization, by definition.

A.4.2. Focus on Winners

It is the nature of researchers to study winners rather than losers. The result is that the emphasis of most articles and books is on the actions necessary to become a winner. However, discussions with successful leaders and organizations indicate that losing is an important part of the nurturing process. They are quick to describe major failures in their pasts and invariably assert that their failures served an important part of their learning process.

A.5 COMPETENCIES AND BEST PRACTICES LIST

Table A.1 presents competencies and best practices listed according to the researchers who analyzed them, where available weighted values are also provided.

TABLE A.1 Project Organization Competencies Listed by Research Source

Competency	Best practice	Value		Source
Project Organization				
Project organization, teams	Multifunctional teams.	1/10	+0.87|+.15	Eisenhardt and Tabrizi, 1990
Project organization	Overlapping development.	2/10	+0.38|−.14	
Project organization, supplier	Supplier involvement.	6/10	+0.08|−.19	
Project organization	Schedule rewards.	7/10	−0.01 −.69	
Project organization	Client acceptance.	5/10	.4101	Lewis, 1995
Project organization	Personnel.	8/10	.3593	
Project organization	Clear statement of requirements.	3/10	13.0	The Standish Group, 1995
Project manager	Proper planning.	4/10	9.6%	
Project organization	Realistic expectations.	5/10	8.2%	
Project organization	Competent staff.	7/10	7.2%	
Project organization	Hard-working, focused staff.	10/10	2.4%	
Project organization	Physical proximity. Physical proximity as measured by desk distance in engineering groups is directly related to positive communication of scientific and technical knowledge.			Allen, 1977
Project organization	Cross-functional teams. More functions result in better communications. Team members tend to communicate more with outsiders who have similar functional backgrounds. Consequently, the more functions on the team, the broader and more comprehensive the external communications. As a result external management rates the performance of the team higher.			Ancona and Caldwell, 1992b

Project organization	Cross-functional teams. The project team is a critical element in achieving project objectives. Cross-functional teams are critical to success. The functional diversity of the cross functional team increases the amount and variety of information available to the team. Downstream problems are identified sooner. Problems identified early on are smaller, less expensive, and easier to fix.	Brown and Eisenhardt, 1995
Project organization	Team tenure. Team tenure plays a role in project success. Teams with long tenure (over three years) tend to become inwardly focused and neglect external communications.	
Project organization	Supplier involvement. Suppliers and customers are key players in the project process. Supplier involvement can alert the team to potential down stream problems early on when they are easier to fix.	Clark et. al., 1987
Project organization	Early involvement of suppliers in the design process.	
Project organization	Multifunctional teams.	
Project organization	Development phases are overlapped.	
Project organization	Overlapping product development phases such as integrating die design and die making significantly reduce product development cycle time.	Clark and Fujimoto, 1991

TABLE A.1 Continued

Competency	Best practice	Value	Source
Project organization	Cross-functional groups.		Cordero, 1991
Project organization	Selecting faster project strategies.		
Project organization	Implementing faster project development strategies.		
Project organization	Managing human resources for speed.		
Project organization	Reduce time to market by shortening each step in the development process.		
Project organization	Rewards have an important effect on project speed. It is particularly true when the project process is predictable. Speed increases because team efforts focus on tasks directly related to the project goal.		
Project organization	Cross-functional teams.		
Project organization	Project planning and control techniques. One objective is to use the most simple methodology necessary to get the job done. Good planning and contol simplifies the management of the product development process.		
Project organization	Resource Intensive techniques. On resource intensive projects, time can be saved by adding cost. If the project is labor intensive, overtime utilization or personnel additions can reduce time to market. One problem with this approach is that often cost increases much more rapidly than time is reduced.		

Project organization	Cross-functional teams. On the most successful projects, cross-functional personnel combine their views in a more highly interactive fashion. Information content is increased.	Dougherty, 1992
Project organization	Rewarding project teams for achieving clear deadlines synchronizes team energies.	Gersick, 1988
Project organization	Buying, licensing and contracting technology obtained from external sources.	Gold, 1987
Project organization	Increased rewards for successful R&D performance. Shorten development times by rewarding designers for speed.	
Project organization	Encouraging internal competition.	
Project organization	Utilizing peer reviews to accelerate progress.	
Project organization	Delegate project steps to suppliers. Supplier involvement during development reduces the workload on the primary team.	
Project organization	Downstream problems specific to functional areas are observed earlier when multifunctional teams are used.	
Project organization	Early involvement of functional groups. 42% of respondents stated that early involvement of the *cross-functional* teams is required. Cooperation of the functional groups is improved. Early involvement helps define product requirements before large amounts of money have been spent.	Gupta and Wileman, 1990

Table **A.1** Continued

Competency	Best practice	Value	Source
Project organization	Cross-functional project teams.		
Project organization	Strong project teams eliminate weak matrix projects.		
Project organization	Availability of new product development resources. A successful project requires good people, facilities, tools, and other resources.		
Project organization	Rewards have an important effect on project speed. It is particularly true when the project process is predictable.		
Project organization	Methodology impact. Performing all the design activities compared to engaging in few correlated positively with a higher new product success rate.		Hise et al., 1989
Project organization	Audits—measure the cost of savings.		Ingram, 1998
Project organization	Emphasize multifunctional training.		Lorenz, 1990
Project organization	Cross-functional teams.		
Project organization	Overlap of project phases.		
Host organization	Senior management engages in subtle control.		
Project organization	Make extensive use of supplier networks.		
Project organization	Multifunctional teams.		
Project organization	The fastest moving projects actively involve suppliers in the process.		Mabert et al., 1992
Project organization	Cross-functional views are a key component of project success.		Myers and Marquis, 1969

Project organization	Overlapping every phase of the development process.			Nonaka, 1990
Project organization	Bringing in representatives in the early stages of the project improves understanding of goals and objectives.			Pelz and Andrews, 1966
Project organization	Optimum group longevity is between three and four years.			Smith, 1979
Project organization	Optimum group longevity is between three and four years.			
Project organization	Overlap development steps.			Stalk and Hout, 1990
Project organization	Involving multifunctional representatives early in the process reduces wait time between steps.			
Project organization	Overlap generations of the project or product development. Basically before all the work on one phase of the project is complete, the next phase is underway.			Zangwill, 1992
Project organization	Cross-functional teams.			
Project organization	Methodologies. Project teams rely on formalized methods, procedures and documents.			
Project organization	Audits.			
Host Organization				
Host organization product	Introduce a unique but superior product.	1/44	0.859	Cooper, 1979
Host organization product	Have a mechanically, technically complex product.	2/44	0.875	
Host organization, market research	Understand buyer behavior.	6/44	0.761	

TABLE A.1 Continued

Competency	Best practice	Value	Source
Host organization market research	Understand size of potential market.	7/44 0.741	
Host organization market skill	Market knowledge and marketing proficiency.	8/44 0.730	
Host organization, technical synergy	Technical and production synergy and proficiency.	9/44 0.680	
Host organization, strategy	Undertake preliminary assessment well.	10/44 0.691	
Host organization, marketing	Strong marketing communications and launch effort.	14/44 0.517	
Host organization, marketing	Have a market derived idea with considerable investment involved.	15/44 0.510	
Host organization, marketing	Undertake preliminary market assessment well.	16/44 0.470	
Host organization, marketing	Undertake market study well.	17/44 0.570	
Host organization, marketing	Understand customer's needs, wants.	18/44 0.583	
Host organization, marketing	Understand buyer price sensitivity.	19/44 0.612	
Host organization, marketing	Understand competitive situation.	21/44 0.616	
Host organization, sales	Have a strong sales force launch effort.	22/44 0.408	
Host organization, sales	Sales force effort is well targeted.	23/44 0.507	
Host organization, product	Highly innovative product, new to market.	24/44 0.449	
Host organization, product	Product lets customer reduce costs.	25/44 0.410	
Host organization, product	Product does unique task for the customer.	26/44 0.564	
Host organization, product	Product is higher quality than competition.	27/44 0.691	
Host organization, finance	Have adequate financial resources.	28/44 0.576	
Host organization, marketing	Have necessary market research resources.	29/44 0.677	

				Cooper and Kleinschmidt, 1987
Host organization, marketing	Customers have great need for the product type.	30/44	0.558	
Host organization advertising	Have a strong adverting/promotion launch effort.	33/44	0.762	
Host organization advertising	Advertising effort is well targeted.	34/44	0.668	
Host organization advertising	Undertake market launch well.	35/44	0.457	
Host organization, marketing	The product is clearly defined by marketplace.	36/44	0.689	
Host organization, technical	The technical solution is clear at start.	37/44	0.577	
Host organization, first to market	Company is the first into market with product.	39/44	0.616	
Host organization, strategy	Undertake initial idea screening well.	40	0.401	
Host organization, marketing	Undertake preliminary market assessment well.	41/44	0.374	
Host organization, technical	Undertake preliminary technical assessment well.	42/44	0.336	
Host organization, marketing	Undertake market study well.	43/44	0.284	
Host organization, finance	Undertake financial analysis well.	44/44	0.338	
Host organization, product	The product concept—what the product will be and do—is well defined.	1/22	0.95	
Host organization, market	The customer's needs, wants, and preferences are well defined.	2/22	0.94	
Host organization, product	The product specifications and requirements are well defined.	3/22	0.94	
Host organization market	The target market is well defined.	4/22	0.92	
Host organization product	The product offers unique benefits to the customer—benefits not found in competitive products.	5/22	0.82	

TABLE A.1 Continued

Competency	Best practice	Value		Source
Host organization, senior management	There is top management guidance/direction for the project.	6/22	0.76	
Host organization product	The product is superior to competition in the eyes of the customer.	7/22	0.73	
Host organization, senior managment	Top management is involved in the day-to-day management of the project.	8/22	0.73	
Host organization, product	The product is innovative—it is the first of its kind in the market.	9/22	0.66	
Host organization, product	The product solves a problem the customer has had with competitive products	10/22	0.64	
Host organization, superior product	The product is higher quality than competitive products.	11/22	0.59	
Host organization senior management	Top management is commited to the project.	12/22	0.52	
Host organization, strategy	Initial screening is proficient.	13/22	0.46	
Host organization, market	Preliminary market assessment is proficient.	14/22	0.46	
Host organization finance	Business and financial analysis is proficient.	17/22	0.37	
Host organization market research	Detailed market study/market research is proficient.	21/22	0.25	
Host organization, technical assessment	Preliminary technical assessment is proficient.	22/22	0.22	
Host organization, technical assessment	Preliminary technical assessment is proficient.	1/6	6.47	Dwyer and Mellor, 1991

Category	Description			Reference
Host organization market assessment	Test marketing is proficient.	2/6	6.40	
Host organization, market assessment	There is a detailed market study.	4/6	6.00	
Host organization, strategy	Initial project screening is proficient.	6/6	5.59	Lewis, 1995
Host organization, senior support	Management support.	9/10	0.3735	Standish Group, 1995
Host organization, senior support	Executive management support.	2/10	13.9%	
Host organization, own-ership	Ownership support.	8/10	5.3%	Clark and Fujimoto, 1991
Host organization	Senior management exercises subtle control of the team and transmits the organizational vision through the heavyweight team leader.			
Host organization	Product is consistent with the organizational image.			
Host organization	Senior management clearly communicates the concept of product integrity, i.e., the vision of the products intended image, performance, and fit with organizational competencies and market needs.			
Host organization	Having a unique, superior product in the eyes of the customer. The product has a real advantage over competition in the market.			Cooper, 1980

TABLE **A.1** Continued

Competency	Best practice	Value	Source
Host organization, market	Having strong market knowledge and market inputs. Performing the market research skillfully.		
Host organization, synergy	Having technological and production synergy and proficiency.		
Host organization, strategy	Making product time to completion a central objective of the firm.		Cordero, 1991
Host organization	Market analysis. Failed new products are better mousetraps that no one wants. Typically they are products conceived and developed in the absence of market information and with no clearly defined market need.		Calantone and Cooper, 1977
Host organization	Senior management support. Dealing with the various challenges faced by major projects necessitates involvement of senior management support. It gives the project credibility and a sense of priority within the organization. Destructive conflicts are minimized.		Gupta and Wileman, 1990

Host organization	Senior management. Senior management plays an important role in attaining project success. Senior management support includes provision of financial, political, and material resources to the team. Senior management gains project approval to go ahead. Subtle control is important to assure superior project performance. Senior management and project leadership work closely together to communicate vision.	Brown and Eisenhardt, 1995
Host organization	In-depth knowledge of the market, customers	Maidique and Zirger, 1985
Host organization	Introduces a product with a high performance to cost ratio	
Host organization	Significant resource commitments to selling and promoting the product.	
Host organization	The product results in a high profit contribution to the organization.	
Host organization	Market introduction of the product is early.	
Host organization	The strengths of the firm enhance the markets and technologies of the project.	
Host organization	Management displays a high level of support for the project.	
Project organization	Multifunctional teams, i.e., smooth execution of all phases of the development process by well coordinated functional teams.	

TABLE **A.1** Continued

Competency	Best practice	Value		Source
Host organization	Product factors. Enhanced technical performance, low cost, reliability, quality, and uniqueness.			
Host organization	Top management commitment.			
Host organization	Build projects on existing corporate strengths.			
Host organization	Market vs. technical. Project success if largely dependent upon market issues rather than purely technical ones. Only 21 percent of successful innovations studied are the result of technology push.			Myers and Marquis, 1969
Host organization	Identifying and understanding user needs was more important than pushing a new technology.			Myers and Marquis, 1969
Host organization	Senior leadership.			Rothewell et al., 1974
Host organization	Senior Management Support.			Zangwill, 1992
The External Environment				
External environment, product competitive	Superior to competing products in meeting customer needs.	3/44	0.832	Cooper, 1979
External environment, product high technology	Have a high technology product.	4/44	0.820	
External environment, product unique	Product has unique features for the customer.	5/44	0.799	
External environment	Being in a large, high need, growth market.	12/44	0.610	

External environment, product competitive	Introducing a low priced product with economic advantages.	13/44	0.578	
External environment, demand	There is customer need for product type.	15/22	0.42	Cooper and Kleinschmidt, 1987
External environment, customer	The product reduces customer costs.	16/22	0.41	
External environment, customer	The product is important to the customer.	18/22	0.32	
External environment, large market	There is large market size.	19/22	0.31	
External environment, market growth	There is rapid market growth.	20/22	0.31	
External environment, market size	Market size (dollar volume) is large.	31/44	0.570	
External environment growth	There is a high growth market.	32/44	0.634	Ancona and Caldwell, 1990
External environment scan competition	Find out what competing firms or groups are doing on similar projects.	1/21	0.791	
External environment	Market factors also impact product success. Early entry into large growing markets, for example.			Maidique and Zirger, 1984/5
External environment	Knowledge of the environment is a key element affecting leader success.			Pfeffer and Salancik, 1978
External environment	Knowledge of the environment. More than any other single factor, the environment impacts organizational structure and the mode of internal processes. Knowledge of the environment is a key element affecting leader success.			Daft et al., 1988

Appendix B

Project Charter Checklist

The project charter checklist is a listing of topics necessary to make a decision regarding the selection and initiation of projects. The subjects listed are generic to many large projects. It usually isn't necessary to expand and explain every topic area listed. The project manager can use the list to selectively develop topics appropriate to the project at hand by detailing those items that apply to your proposed project.

1. **Signature page**. List all stakeholders required to approve the project. Include a one-paragraph description of the project and statement about the action being approved.
2. **Executive summary** (one page or less).
3. **Overview and description.**
   3.1. Project product or service.
       3.1.1. Deliverables.
       3.1.2. Initial work breakdown structure.
       3.1.3. Project scope: Summarize the elements of the project that involve the expenditure of funds.
       3.1.4. Functional requirements of the project. What does the customer want it to do?

3.1.5.   Technical requirements of the project. How will it be done?

3.1.6.   Uniqueness of the project's product of service.

3.1.7.   Factors that differentiate the product or service from competition.

3.1.8.   Breadth of the product or service line.

3.1.9.   Price competitiveness.

3.1.10.   Substitute products.

3.1.11.   Product service image, reputation, quality.

3.2.   Need for the project.

3.3.   History.

3.3.1.   Founders, initiators, or owners of the project.

3.3.2.   History of the project.

3.3.3.   Nature of the project. Has the nature of the project changed or evolved since inception? Is it intended to place emphasis on different areas?

3.4.   Plans for future expansion or change.

3.5.   Alternatives investigated.

3.6.   Compatibility with the host organization's strategic plan.

3.6.1.   Mission and vision of the project. How does it support the host organization strategy.

3.6.2.   SWOT (strengths, weaknesses, opportunities, threats) analysis of the project related to the host organization.

3.7.   Project goals and strategies.

3.7.1.   Long- and short-term goals.

3.7.2.   Functional goals, deliverables, measurements of success, and critical success factors.

3.8.   Key success factors.

3.9.   Performance measurement metrics.

3.10.   Stakeholders involved in and affected by the project. Description of stakeholder issues.

3.11.   Technology.

3.11.1.   Patents, trade names, and intellectual property aspects of the product or service.

3.12.2.   Research and development activities associated with the project.

3.12.3.   Technology trends on which technologies does the project focus? State-of-the-art necessary for products in each project?

4. Selection and evaluation.

4.1.   Financial return and portfolio ranking. Expected product or service financial performance.

4.1.1. Ranking methods used: net present value, internal rate of return, payback?

4.1.2. Cash flow Projections.

    4.1.2.1. Detail by month for first year.

    4.1.2.2. Detail by quarter for second and third years.

    4.1.2.3. Notes of explanation.

4.1.3. Sources and applications of funding.

4.1.4. Historical financial reports for projects already in existence.

    4.1.4.1. Balance sheets, income statements, cash flows for past three years.

    4.1.4.2. Explanation and notes explaining the statements.

4.1.5. Abnormal, nonrecurring, or unusual items in the financials.

4.1.6. Guarantees, warranties, litigation, etc.

4.1.7. Inventory calculation method. Control systems.

4.1.8. Depreciation methods, depletion, amortization. Which items are capitalized and which expensed? Any deferred write-offs?

4.1.9. Accounting methods similar to the rest of the industry?

4.1.10. Debt description. Does the project have any long term or short term debt, secured or unsecured, or has the project guaranteed such debt on the behalf of others?

4.1.11. Cost of capital relative to the industry.

4.1.12. Leverage position. Borrowed money and fixed assets vs. expected return.

4.1.13. Cost control effectiveness.

4.1.14. Describe all bank relationships and credit lines.

4.1.15. Has the project made (a) any private placements of its equity or debt securities or (b) any public sale of its equity or debt securities? If so, furnish complete details.

4.1.16. Claims and litigation. Identify parties, amount involved, the names of the involved, and copies of all documents with respect thereto.

4.1.17. Describe insurance coverage: plant, equipment, properties, work interruption, key employees, other.

4.1.18. Any unusual contracts relating to the project, its products or services.

4.1.19. Resource requirements. Resources needed from the host organization functional areas.

    4.1.19.1. Use of sub contractors and vendors.

    4.1.19.2. Major equipment requirements.

4.1.20. Facilities.

    4.1.20.1. Location and age.

    4.1.20.2. Capabilities. Capacities?

    4.1.20.3. Plans for expansion and growth.

    4.1.20.4. Capital Equipment list.

    4.1.20.5. Excess capacity?

    4.1.20.6. Facilities layout.

    4.1.20.7. Description of all offices, plants, laboratories, warehouses, stores, outlets, or other facilities including size of plot, square footage of enclosed space, etc.

4.1.21. Personnel resources.

    4.1.21.1. Management and leadership resources. Describe characteristics of management, including any of the following that are pertinent: age, education, compensation, past project experience, special distinctions, employment contracts, bonus and profit sharing plans, other employee fringe benefits, pension plans.

    4.1.21.2. Team personnel.

        4.1.21.2.1. Number of team members needed.

        4.1.21.2.2. If dependent upon technology, give details of specialists, i.e., number of doctorates, engineers, specialists, technicians, etc.

        4.1.21.2.3. Description of labor relations, complaints, and costs. Any problems in obtaining personnel? Any turnover problems?

        4.1.21.2.4. Employee skill levels.

        4.1.21.2.5. Incentives used to motivate employees.

        4.1.21.2.6. Ability to level peaks and valleys of personnel usage on the project.

        4.1.21.2.7. Specialized skills.

4.1.21.2.8. Employee experience.

4.1.22 Suppliers and vendors. Names and addresses. Volume of purchases. Are other sources readily available? Dependent upon any one supplier? Any long-term contracts?

4.1.23. Sub contractors? Describe work done and availability of alternative contractors. Any contracts?

4.2. Risk analysis.

4.2.1. Risk related to the portfolio of projects.

4.2.2. Project specific risks (covered under project planning, implementation, and control).

4.2.3. Subjective factors to consider.

4.3.1. Fit with existing services or product lines.

4.3.2. Constraints.

4.3.3. Assumptions.

4.3.4. Additional project or investigation required.

4.3.5. Market and strategic positioning.

4.3.5.1. What geographical area does the project serve? Are there any limitations on what markets can be reached, i.e., freight, duties, service, maintenance, tariffs, government regulations, etc.?

4.3.5.2. Major competitors. Nature and area of competition. Direct or indirect? What is the degree of competition? Can new companies easily enter the field? Do the project's competitors have greater resources? Are they longer established and better recognized? Ease of competitive entry?

4.3.5.3. Describe your project's projection of sales and earnings for the next three years, including explanations with respect to any increase or decrease.

4.3.5.4. Market share expected. Is it a new technology? Is market positioning already established?

4.3.5.5. Effectiveness of sales distribution.

4.3.5.6. Advertising and promotion effectiveness.

4.3.5.7. Are sales concentrated in a few products or to a few customers?

4.3.5.8. Sales organization effectiveness. Knowledge of customer needs.

4.3.5.9. Pricing strategy and pricing flexibility.

4.3.6. Public image.

    4.3.6.1. Articles, press releases, employee interviews.

    4.3.6.2. Project self-image and personality.

    4.3.6.3. Claims and litigation. Identify parties, amount involved, the names of the involved, and copies of all documents with respect thereto.

    4.3.6.4. Any unusual contracts relating to the project, its products or services.

5. The project plan.

5.1. Time schedule for project completion.

    5.1.1. Time measurement metrics. Milestones, decision points, total project completion time.

    5.1.2. Critical path identified.

5.2. Budget and cash flow.

5.3. Risk analysis. Project specific.

5.4. Quality plan.

5.5. Communications plan.

5.6. Project control, variance reporting structure and procedures, measurement metrics.

5.7. Scope management and control.

5.8. Relationship with suppliers and customers.

5.9. Information systems.

    5.9.1. Timeliness and accuracy of variance information.

    5.9.2. Relevance of information for tactical decisions.

    5.9.3. Ability to use information that is provided.

6. Supporting documents.

6.1. Personal resumes, letters of reference, job descriptions, employment agreements, letters of intent, copies of leases, contracts, legal documents.

6.2. Brochures, catalogues, mailers, publicity releases, newspaper or magazine articles, literature, and the like distributed by the project or concerning the project, its products, personnel, or services.

6.3. Copies of pertinent contracts.

6.4. Other information and materials necessary for a complete presentation.

Appendix C

Risk Analysis Matrix: Key Risk Factors Affecting Project Success

Listed herein are a series of statements about the project. Research has shown that these are the critical variables that represent project risk. Read each statement and evaluate the degree or probability of occurrence of the risk associated with the statement using a ranking of high risk, medium risk, or low risk. Then assess the impact if the risk event occurs using a ranking of high impact, medium impact, or low impact. For statements that you feel represent significant risk and impact, describe the response planned (e.g., avoidance, mitigation, deflection).

OVERALL RISK FACTORS

	Probability (H, M, L)	Impact (H, M, L)	Risk strategy
The overall evaluation of the project is that there is a high potential for failure.	_____	_____	_____
The size of the risk is such that project failure could cripple or destroy the organization.	_____	_____	_____

323

PROJECT OUTPUT (PRODUCT OR SERVICE) RISK FACTORS

	Probability (H, M, L)	Impact (H, M, L)	Risk strategy
The project product or service:			
Is the same as competition.	_____	_____	_____
Differs from the company's core competence.	_____	_____	_____
Concept is unclear and hard to understand.	_____	_____	_____
Offers no intrinsic value to the customer.	_____	_____	_____
Offers no innovative features.	_____	_____	_____
Solves no customer problems.	_____	_____	_____
Is commonplace.	_____	_____	_____
Is inconsistent with the project image.	_____	_____	_____
Compared to competition, the project product or service:			
Is clearly inferior.			
Has inferior technical performance.	_____	_____	_____
Has higher cost.	_____	_____	_____
Is less reliable.	_____	_____	_____
Has poorer quality.	_____	_____	_____

MARKET RISK FACTORS

	Probability (H, M, L)	Impact (H, M, L)	Risk strategy
The market for the project product or service:			
Is unattractive.	_____	_____	_____
Is highly competitive.	_____	_____	_____
Is declining.	_____	_____	_____
Has satisfied customer demand.	_____	_____	_____
The project output represents late entry into the market.	_____	_____	_____
Customers are not involved.	_____	_____	_____
Suppliers are not involved.	_____	_____	_____
Users needs are not identified nor understood.	_____	_____	_____

INTERNAL ORGANIZATION RISK FACTORS

	Probability (H, M, L)	Impact (H, M, L)	Risk strategy
Project has visible, top management lack of support.	_____	_____	_____
Project is not built on existing project strengths.	_____	_____	_____
Project has inadequate resources (time, labor, materials, or money) available.	_____	_____	_____
Senior management is not working closely with project leadership to develop the project concept.	_____	_____	_____
Project is incompatible with project culture.	_____	_____	_____

TEAM RISK FACTORS

	Probability (H, M, L)	Impact (H, M, L)	Risk strategy
Members do not represent all functional areas affected by the project.	_____	_____	_____
The project team:			
Execution is disorganized.	_____	_____	_____
Does not work as a team to define goals, develop workable plans, prioritize work Infrequently communicates with outsiders.	_____	_____	_____
Lacks adequate training in multifunctional areas pertinent to the project.	_____	_____	_____
Does not resolve conflicts.	_____	_____	_____
Does not resolve conflicts at lower levels.	_____	_____	_____
Has a limited degree of internal communication.	_____	_____	_____

PROJECT LEADER RISK FACTORS

	Probability (H, M, L)	Impact (H, M, L)	Risk strategy
The project leader:			
Has a history of poor performance on projects.	_____	_____	_____

Has limited decisionmak-
ing accountability.   _____   _____   _____
Is a poor politician who
fails to lobby for product
support, does not en-
sure resources, and
does not buffer the team
from outside pressure.   _____   _____   _____
Discourages team commu-
nications with outsiders.   _____   _____   _____
Gives team members little
freedom to work autono-
mously within the con-
straints of the project vi-
sion.   _____   _____   _____
Is a poor manager of the
group.   _____   _____   _____
Is a lower level manager
with little authority.   _____   _____   _____
Fails to attract top team
members to the group.   _____   _____   _____
Inadequately communi-
cates a clear vision
of the project objec-
tives and goals to the
team.   _____   _____   _____
Ineffectually gathers, facili-
tates, and translates ex-
ternal information for
team members.   _____   _____   _____

PLANNING RISK FACTORS

	Probability (H, M, L)	Impact (H, M, L)	Risk strategy
The project:			
Is expected to lose money.	_____	_____	_____
Has infrequent or no milestones.	_____	_____	_____
Has an insufficient degree of planning.	_____	_____	_____
Target market is undefined.	_____	_____	_____
Risks have been unsatisfactorily identified and quantified and lack developed responses.	_____	_____	_____

Critical success variables have not been identified. _____ _____ _____

Supplier network is not integrated into the development process. _____ _____ _____

Specifications are not thorough, complete nor clearly reflect customer needs. _____ _____ _____

Concept is unclear and difficult to understand. _____ _____ _____

Technology is changing as the project is underway and the customer wants the latest technology. _____ _____ _____

Customer does not understand the project technology. _____ _____ _____

CONCLUSION

_____ Proceed
_____ Proceed with reservations
_____ Resolve problems before proceeding
_____ Abandon

Explanation of Conclusion

_____
_____
_____

References

Sources from the fields of leadership, group psychology and behavior, team performance, and project management.

Aaker, D. A. 1983. Organizing a strategic information scanning system. *California Management Review* 25(2):76–83.

Alderfer, C. P. 1976. Boundary relations in organizational diagnosis. In: M. Meltzer and F. Wickert, eds., *Humanizing Organizational Behavior.* Charles C. Thomas, Springfield, IL: pp. 145–175.

Allen, T. J. 1970. Communications networks in R&D labs. *R&D Management* 1:14–21.

Allen, T. J. 1971. Communications, technology transfer, and the role of technical gatekeeper. *R&D Management* 1:14–21.

Allen, T. J. 1977. *Managing the Flow of Technology: Technology Transfer and the Dissemination of Technological Information with the R&D Organization.* MIT Press, Cambridge, MA.

Allen, T. J., and Cohen, S. 1969. Information flow in R&D laboratories. *Administrative Science Quarterly* 14: 12–19.

Allen, T. J., Lee, D., and Tushman, M. 1980. R&D performance as a function of internal communication, project management, and the nature of work. *IEEE Transactions on Engineering Management* 27:2–12.

Ancona, D. G., and Caldwell, D. F. 1990. Beyond boundary spanning: managing external dependence in product development teams. *Journal of High Technology Management Research* 1:119–135.

Ancona, D. G., and Caldwell, D. F. 1992a. Demography and design: predictors of new product team performance. *Organization Science* 3:321–341.

Ancona, D. G., and Caldwell, D. F. 1992b. Bridging the boundary: external process and performance in organizational teams. *Administrative Science Quarterly* 37:634–665.

Ansoff, H. I. 1975. Managing strategic surprise by response to weak signals. *California Management Review* 18(2): 21–33.

Argenti, J. 1976. Corporate Collapse: The Cases and Symptoms. McGraw-Hill, New York.

Avolio, B. J., Atwater, L., and Lau, A. 1993. A multi-rater view of transformational leadership behavior: key predictors of army camp performance. Paper delivered at the Academy of Management Meeting, Atlanta, GA.

Banner, D., and Gagnle, T. E. 1995. *Designing Effective Organizations: Traditional and Transformational Views.* Sage, Newbury Park, CA.

Barney, J., and Hansen, M. 1994. Trustworthiness as a source of competitive advantage. *Strategic Management Journal* 15.

Bass, B. M. 1961. Some aspects of attempted, successful, and effective leadership. *Journal of Applied Psychology* 45(2):120–122.

Bass, B. M. 1990. *Bass & Stogdill's Handbook of Leadership: Theory, Research and Applications.* The Free Press, New York.

Bass, B., and Avolio, B. 1990. *The Multi-Factor Leadership Questionnaire.* Consulting Psychologists Press, Palo Alto, CA.

Bassiry, G. R., and Dekmejian, R. H. 1990. The American corporate elite. *Business Horizons* 33(3).

Benne, K. D., and Sheats, P. 1948. Functional roles of group members. *Journal of Social Issues* 2:42–47.

Bhattacharya, R., Devinney, T., and Pitulla, M. 1998. A formal model of trust based on outcomes. *Academy of Management Review* 23(3).

Bibeault, D. B. 1982. *Corporate Turnaround: How Managers Turn Losers into Winners.* McGraw-Hill, New York.

Boeker, W. 1992. Power and management dismissal: scapegoating at the top. *Administrative Science Quarterly* 37(3):400–421.

Bourgeois, L. J. 1980. Strategy and environment: a conceptual integration. *Academy of Management Review* 5:25–39.

Braham, J. 1991. Engineering your way to the top. *Machine Design* 63(7).

Brown, S. L., and Eisenhardt, K. M. 1995. Product development: past research, present findings, and future directions. *Academy of Management Review* 20.

Cabanis-Brewin, J. 2000. *Best Practices Report* 1.

Calantone, R., and Cooper, R. G. 1977. A typology of industrial new product failure. In: Greenberg and Bellinger, eds., *Contemporary Marketing Thought.* American Marketing Association, Chicago.

Calantone, R., and Cooper, R. G. 1981. New product scenarios: prospects for success. *Journal of Marketing* 45:48–60.

Campion, M., Papper, E. M., and Medsker, G. J. 1996. Relations between work team characteristics and effectiveness: replication and extension. *Personnel Psychology* 49:429–452.

Cascio, Wayne F. 2000. Managing a Virtual Workplace. *Academy of Management Executive* 14:(3).

Castiglione, B. 1993. *The Book of the Courtier 1528.* As reprinted by the Hartwick Humanities in Management Institute. 1993.

Chandy, P. R. 1991. Chief executive officers: their backgrounds and predictions for the 90's. *Business Forum* 16(1):18–19.

Child, J. 1977. Management. In Parker, S. R., Brown, R. K., Child, J., and Smith, M. A., eds., *The Sociology of Industry*, 3rd ed. Allen and Unwin, London.

Clark, K. B., Chew, W. B., and Fujimoto, T. 1987. Product development in the world auto industry. *Brookings Papers on Economic Activity* 3:729–781.

Clark, K. B., and Fujimoto, T. 1991. *Product Development Performance.* Harvard Business School Press, Boston.

Clark, K. B., Hayes, R. H., and Lorenz, C. 1985. *The Uneasy Alliance: Managing the Productivity Technology Dilemma.* Harvard Business School Press, Boston, pp. 337–375.

Cleland, D. 1999. *Project Management Strategic Design and Implementation.* McGraw-Hill, New York.

Cooper, R. G. 1975. Why new industrial products fail. *Industrial Marketing Management,* 4:315–326.

Cooper, R. G. 1979. The dimensions of industrial new product success and failure. *Journal of Marketing* 43:93–103.

Cooper, R. G., and Kleinschmidt, E. J. 1987. New products: what separates the winners from losers? *Journal of Product Innovation Management* 10.

Cooper, R. G. 1980. Project New Products: factors in new product success. *European Journal of Marketing* 14: 277–292.

Cordero, R. 1991. Managing for speed to avoid product obsolescence: a survey of techniques. *Journal of Product Innovation Management* 8:283–294.

Crawford, C. M. 1979. New product failure rates—facts and fallacies. *Research Management* September: 9–13.

Culnan, M. J. 1983. Environmental scanning: the effects of task complexity and source accessibility on information gathering behavior. *Decision Sciences* 14:194–206.

Daft, R. L., Sormunen, J., and Parks, D. 1988. Chief executive scanning, environmental characteristics, and company performance: an empirical study. *Strategic Management Journal* 9:123–139.

DeMatteo, J. S., Eby, and Sundstrom, E. 1998. Team based rewards: current empirical evidence and directions for future research. *Research in Organizational Behavior* 20.

Dess, G. G., and Robinson, R. B. 1989. Measuring Organizational performance in the absence of objective measures. *Strategic Management Journal* 5:265–273.

Dewhirst, H. Dudley, Arvey, Richard D., and Brown, E. M. 1978. Satisfaction and performance in research and development tasks as related to information accessibility. *IEEE Transactions on Engineering Management* 25:58–63.

Dill, W. R. 1958, Environment as an influence on managerial autonomy. *Administrative Science Quarterly* 3:409–443.

Doney, P., Cannon, J., and Mullen, M. 1998. Understanding the influence of national culture on the development of trust. *Academy of Management Review* 23(3).

Dougherty, D. 1990. Understanding new markets for new products. *Strategic Management Journal* 11:59–78.

Dougherty, D. 1992. Interpretive barriers to successful product innovation in large firms. *Organization Science* 3:179–202.

Dubin, S. S. 1972. *Professional Obsolescence*. Lexington Books, D.C. Heath, Lexington, MA.

Duncan, R. B. 1972. Characteristics of organizational environments and perceived environmental uncertainty. *Administrative Science Quarterly* 17:313–327.

Dwyer, L., and Mellor, R. 1991. Organizational environment. New product process activities, and project outcomes. *Journal of Product Innovation and Management* 8:39–48.

Eisenhardt, K. M. 1989. Making fast strategic decisions in high velocity environments. *Academy of Management Journal* 32.

Eisenhardt, K. M., and Tabrizi, B. 1990. Accelerating adaptive processes: product innovation in the global computer industry. *Administrative Science Quarterly*.

Elangovan, A. R., and Shapiro, D. 1998. Betrayal of trust in organizations. *Academy of Management Review* 23(3).

Fayol, H. 1916. (Revised by Gray, Irwin. 1984.) *General and Industrial Management*. IEEE Press, New York.

Feldman, M. S., and March, James G. 1981a. Information in organizations as signal and symbol. *Administrative Science Quarterly* 26: 171–186.

Feldman, M. S., and March, James G. 1981b. Organizational factors and individual performance. *Journal of Applied Psychology* 53:86–92.

Finn, R. 1993. A synthesis of current research on management competencies. Working paper, Henley Management College, Brunel University of West London.

Forrester, R., and Drexler, A. B. 1999. A model for team based organization performance. *The Academy of Management Executive* 13(3).

Fortune. 1989. How managers can succeed through speed. February 13, pp. 54–59.

Gadeken, O. C. 1989. DSMC studies program manager competencies. *Program Manager* January-February.

Gadeken, O. C. 1996. Project managers as leaders: competencies of top performers. Paper delivered at Project World.

Galbraith, Jay R. 1973. *Designing Complex Organizations.* Addison Wesley, Reading, MA.

George, A. L. 1980. *Presidential Decisionmaking in Foreign Policy.* Westview, Boulder, CO.

Gersick, C. J. G. 1988. Marking time: predictable transitions in group tasks. *Academy of Management Journal* 31:9–41.

Gersick, C. J. G. 1989. Marking time: predictable transitions in task groups. *Academy of Management Journal* 32:274–309.

Gerstenfeld, A. 1976. A study of successful projects, unsuccessful projects, and projects in process in West Germany. *IEEE Transactions in Engineering Management* 23:116–123.

Gilbreth, L. M. 1924. *The Quest of the One Best Way.* 1990 Edition. Soc. Women Eng., New York.

Globe, S., Levy, G. W., and Schwartz, C. M. 1973. Key factors and events in the innovation process. *Research Management* 16:8–15.

Gold, B. 1987. Approaches to accelerating product and process development. *Journal of Product Innovation Management* 4:81–88.

Govindarajan, V. 1989. Implementing competitive strategies at the business unit level: implications of matching managers to strategies. *Strategic Management Journal* 10.

Griffith, B. C., and Mullins, N. C. 1972. Coherent social groups in scientific change. *Science* 177:959–964.

Grinyer, P. H., and Norburn, D. 1975. Planning for existing markets: perceptions of executives and financial performance. *Journal of the Research Statistical Society* 138(Part 1):70–97.

Gupta, A. K., and Govindarajan, V. 1984. Business unit strategies, managerial characteristics, and business unit effectiveness at strategy implementation. *Academy of Management Journal* 27.

Gupta, A. K., and Wilemon, D. L. 1990. Accelerating the development of technology based new products. *California Management Review* 32(2): 24–44.

Hackman, J. R., ed., 1990. *Groups that Work (and Those that Don't): Creating Conditions for Effective Teamwork.* Josey-Bass, San Francisco.

Hagan, J., and Choe, S. 1998. Trust in Japanese interfirm relations: institutional sanctions matter. *Academy of Management Review* 23(3).

Hambrick, D. C., and Mason, P. A. 1984. Upper echelons: the organization as a reflection of its top managers. *Academy of Management Review* 9.

Hatch, R. S. 1957. Product failures attributed mainly to a lack of testing and faulty marketing. *Industrial Marketing* 43 (February):112–126.

Hayes, R. H., Wheelwright, S. C., and Clark, K. 1988. *Dynamic Manufacturing.* The Free Press: New York.

Hise, R. T., and McDaniel, S. W. 1988. American competitiveness and the leader—who's minding the shop? *Sloan Management Review* 29:49–55.

Hise, R. T., O'Neal, L., Parsuraman, A., and McNeal, J. U. 1989. The effect of product design activities on commercial success levels of new industrial products, *Journal of Product Innovation Management* 6(1)March:43–50.

Hise, R. T., O'Neal, L., Parsuraman, A., and McNeal, J. U. 1990. Marketing/R&D interaction in new product development: implications for new product success rates. *Journal of Product Innovation Management* 7: 142–155.

Hitt, M. A., Ireland, R. D., and Palia, K. A. 1982. Industrial firms' grand strategy and functional importance: moderating effects of technology and uncertainty. *Academy of Management Review* 25.

Holland, W. E., Stead, B. A., and Leibrock, R. C. 1976. Information channel source selection as a correlate of technical uncertainty in a research and development organization. *IEEE Transactions on Engineering Management* 23:163–167.

Hollensbe, E. C., and Guthrie, James P. 2000. Group pay for performance plans: the role of spontaneous goal setting. *Academy of Management Review* 25(4).

Homans, G. 1961. *Social Behavior: Its Elementary Forms.* Harcourt, Brace, New York.

Hopkins, D. S. 1980. *New Product Winners and Losers.* Conference Board Report No. 773. National Industrial Conference Board, New York.

Hornaday, J. A., and Aboud, J. 1971. Characteristics of successful entrepreneurs. *Personnel Psychology* 24:141–153.

Ibbs, C. W., and Kwak, Y-H. 1997. *The Benefits of Project Management: Financial and Organizational Rewards to Corporations.* Project Management Institute Educational Foundation, Upper Darby, PA.

Ingram, T. 1998. *How to Turn Computer Problems into Competitive Advantage.* Project Management Institute, Upper Darby, PA.

Ireland, L. R. 1991. *Quality Management for Projects and Programs.* Project Management Institute, Upper Darby, PA.

Janis, I. L. 1982. *Groupthink.* Houghton Mifflin, Boston.

Janis, I. L. 1985. Sources of error in strategic decision making. In: Johanes M. Pennings and Associates, eds. *Organizational Strategy and Change.* Josey-Bass, San Francisco, pp. 157–197.

Karau, S. J., and Williams, K. D. 1993. Social loafing: meta-analytic review and theoretical integration. *Journal of Personality and Social Psychology* 65.

Katz, R., and Tushman, M. L. 1979. Communication patterns, project performance and task characteristics: an empirical evaluation and integration in an R&D setting. *Organizational Behavior and Human Performance* 23:139–162.

Katz, R., and Tushman, M. L. 1981. An investigation into the managerial roles and career paths of gatekeepers and project supervisors in a major R&D facility. *R&D Management* 11:103–110.

Katz, R. 1982. The effects of group longevity on project communication and performance. *Administrative Science Quarterly* 27 March: 81–104.

Keller, R. T. 1986. Predictors of the performance of project groups in R&D organizations. *Academy of Management Journal* 29:715–726.

Keller, R. T., and Holland, W. E. 1983. Communicators and innovators in research and development organizations. *Academy of Management Journal* 26:742–749.

Kingdon, O. R. 1973. *Matrix Organization: Managing Information Technologies.* Tavistock, London.

Kirkpatrick, Shelley A. and Locke, E. A. 1991. Leadership: do traits matter? *Academy of Management Executive* 5:48–60.

Knight, K. 1977. *Matrix Management.* Gower, London.

Knoll, K., and Jarvenpaa, S. L. 1998. Working together in virtual teams. In: Igbaria and Tan, eds., *The Virtual Workplace.* Idea Group Publishing, Hershey, PA.

Kotter, J. P. 1982. *The General Managers.* McGraw-Hill, New York.

Kouzes, J., and Posner, B. Z. 1993. *Credibility.* Jossey-Bass, San Francisco.

Kramer, R. M., and Tyler, T. R. 1996. *Trust in Organizations: Frontiers of Theory and Research.* Sage Publications, Thousand Oaks, CA.

Lamb, R. B. 1987. *Running American Business: Top CEOs Rethink Their Major Decisions.* Basic Books, New York.

Lawrence, P. R., and Lorsch, J. W. 1967. *Organization and Environment.* Harvard Business School, Boston.

Lazo, H. 1965. Finding a key to success in new product failures, *Industrial Marketing,* 50 (November), 74–77.

Lewicki, R., McAllister, D., and Bies, R. 1998. Trust and distrust: new relationships and realities. *Academy of Management Review* 23(3).

Lieberson, S., and O'Connor, J. F. 1972. Leadership and organizational performance: a study of large organizations. *American Sociological Review* 37:117–130.

Likert, R. 1961. New Patterns of Management. McGraw-Hill, New York.

Likert, R. 1967. The Human Organization. McGraw-Hill, New York.

Lindsay, W. M., and Rue, L. W. 1980. Impact of the organization environment on the long range planning process: a contingency view. *Academy of Management Journal* 23:385–404.

Mabert, V. A., Muth, J. F., and Schmenner, R. W. 1992. Collapsing new product development times: six case studies. *Journal of Product Innovation Management* 9:200–212.

Machiavelli, N. 1520. (Reprinted by The Hartwick Humanities in Management Institute. 1993). *The Prince.* Hartwick College, Oneonta, NY.

Maidique, M. A., and Zirger, B. J. 1984. A study of success and failure in product innovation: the case of the U.S. electronics industry. *IEEE Transactions in Engineering Management* 4:192–203.

Maidique, M. A., and Zirger, B. J. 1985. The new product learning cycle. *Research Policy* 14: 299–203.

Maidique, M., and Zirger, B. J. 1990. A model of new product development: an empirical test. *Management Science* 36.

Makridakis, S. 1991. What we can learn from corporate failure? *Long Range Planning* 24, August.

Marquis, D. G. 1969. The anatomy of successful innovations, *Innovation Magazine* 1 (November):28–37.

Marquis, D. G., and Straight, D. L., 1965. *Organizational Factors in Project Performance.* M.I.T. Sloan School of Management Working Paper No. 1331. Cambridge, MA.

McGrath, J. E. 1984. *Groups: Interaction and performance.* Prentice-Hall, Englewood Cliffs, N.J.

McKnight, D., Cummins, L., and Chervany, N. L. 1998. Initial trust formation in new organizational relationships. *Academy of Management Review* 23(3).

Menzel, H. 1965. Information needs and uses in science and technology. In: *Annual Review of Science and Technology.* Wiley, New York, pp. 41–69.

Meyer, A. D. 1979. Adapting to environmental jolts. *Administrative Science Quarterly* 17:313–327.

Miles, R. E., and Cameron, K. 1982. *Coffins Nails and Corporate Strategies.* Prentice-Hall, Englewood Cliffs, N.J.

Miller, D., Toulouse, J-M. 1986. Chief executive personality and corporate strategy and structure in small firms. *Management Science* 32:1389–1409.

Miller, D., Toulouse, J-M., and Belanger, N. 1985. Top executives' personality and corporate strategy: three tentative types. In *Advances in Strategic Management*, vol. 3. JAI Press, Greenwich, CT.

Millison, M. R., Raj, S. P., and Wilemon, D. 1995. A survey of major approaches for accelerating new product development. *Journal of Product Innovation Management* 9:53–69.

Mohrman, S. A., Cohen, S. G., and Mohrman, A. M. 1995. *Designing Team Based Organizations.* Josey-Bass, San Francisco.

Moretti, D. M., Morken, C. L., and Borkowski, Jeanne M. 1991. Profile of the American leader: comparing *Inc.* and *Fortune* executives. *Journal of Business & Psychology* 6: 193–205.

Morrison, E. W., and Milliken, F. J. 2000. Organizational silence: a barrier to change and development in a pluralistic world. *Academy of Management Review.* 25(4):706–725.

Moyer, C., McGuigan, J. R., and Kretlow, W. J. 1995. *Contemporary Financial Management.* West Publishing Company, Minneapolis/St. Paul.

Myers, S., and M., D. G. 1969. *Successful Industrial Innovations.* NSF 69–17. National Science Foundation, Washington, DC.

National Industrial Conference Board. 1964. Why new products fail. *The Conference Board Record.* NICB, New York.

Ness, J. A., and Cucuzza, T. G., 1995. Tapping the full potential of ABC. *Harvard Business Review* July-August.

Nevins, J. L., and Whitney, D. E. 1992. *Concurrent Design of Products and Processes.* McGraw-Hill, New York.

Nonaka, I. 1990. Redundant, overlapping organizations: a Japanese approach to managing the innovation process. *California Management Review* 32(3):27–38.

Norburn, D. 1989. The chief executive: a breed apart. *Strategic Management Journal* 10.

Nutt, P. C. 1999. Surprising but true: half the decisions in organizations fail. Academy of Management Executive 13:4.

O'Keefe, R. D., Kernaghan, J. A., and Rubenstein, A. H. 1975. Group cohesiveness: a factor in the adoption of innovations among scientific work groups. *Small Group Behavior* 6:282–292.

Pelz, D., and A., Frank M. 1966. *Scientists in Organizations*. Wiley, New York.

Pennypacker, J. 2000. *Best Practices Report.* 1(11) September.

Pfeffer, J. 1981. *Power in Organizations.* Pitman, Marshfield, MA.

Pfeffer, J. 1992. *Managing with power.* Harvard Business Press, Boston.

Pfeffer, J., and Salancik, G. R. 1978. *The External Control of Organizations,* Harper and Row, New York.

Pinto, J. K., and D. P. Slevin. 1988. Project success: definitions and measurement techniques. *Project Management Journal* 19(1): 67–72.

Pinto, J. K., and D. P. Slevin. 1989. Critical success factors across the project life cycle. *Project Management Journal* 67–72.

Porter, L. W., Lawler, E. E., and Hackman, R. J. 1975. *Behavior in Organizations.* McGraw-Hill, New York.

Priem, R. L. 1993. Leader judgment, strategy making process fit, and firm performance. Paper delivered at the Academy of Management, June.

Radosevich, L. 1999. Project management: measuring up. *CIO.* September 15.

Rhyne, L. C. 1985. The relationship of information usage characteristics to planning system sophistication: an empirical examination. *Strategic Management Journal* 6: 319–337.

Roberts, R. W., and Burke, J. E. 1974. Six new products—what made them successful. *Research Management* 16:21–24.

Rogers, E. M., and Shoemaker, F. F. 1971. *Communications of Innovations: A Cross-cultural Approach.* Free Press, New York.

Rogers, R. 1994. *The Psychological Contract of Trust.* Development Dimensions International, Pittsburgh.

Rosenau, M. D. 1988. Speeding your product to market. *Journal of Consumer Marketing* 5:23–40.

Rothwell, R. 1972. Factors for success in industrial innovations. *Project SAPPHO— A Comparative Study of Success and Failure in Industrial Innovation.* S.P.R.U., Brighton, UK.

Rothwell, R., Freeman, C., Horsley, A., Jervis, V.T.P., Robertson, A., and Townsend, J. 1974. SAPPHO updated—project Sappho phase II. *Research Policy* 3:258–291.

Rousseau, D., Sitkin, S., Burt, R., and Camerer, C. 1998. Not so different after all: A cross discipline view of trust. *Academy of Management Review* 23(3).

Rubenstein, A. H., Chakrabarti, A. K., O'Keefe, R. D., Souder, W. E., and Young, H. C. 1976. Factors influencing success at the project level. *Research Management* 16:15–20.

Saint Benedict of Nursia. ca. 1520. (Reprinted by the Hartwick Humanities in Management Institute. 1993). The Rule of St. Benedict. Hartwick College, Oneonta, NY.

Salancik, G. R., and Pfeffer, J. 1978. A social information processing approach to job attitudes and task design. *Administrative Science Quarterly* 23:224–253.

Schmidt, S. M., and Kochan, T. A. 1972. The concept of conflict: toward conceptual clarity. *Administrative Science Quarterly* 17:359–370.

Schnapper, M. 2000. Measure, manage, and magnify performance. *Best Practices Report* 1(11) September.

Schuster, J. R., and Zingheim, P. K. 1992. Building pay environments to facilitate high performance teams. *ACA Journal* 2(1).

Schwartz, K. B., and Menon, K. 1985. Executive succession in failing firms. *Academy of Management Journal* 28(3):686–697.

Shaw, M. E. 1981. *Group Dynamics: The Psychology of Small Group Behavior.* McGraw-Hill, New York.

Shepard, H. A. 1956. Creativity in R&D teams. *Research and Engineering* October:10–13.

Shepphard, B., and Sherman, D. 1998. The grammars of trust: a model and general implications. *Academy of Management Review* 23(3).

Sherif, M. et al. 1961. *Intergroup Conflict and Cooperation: The Robber's Cave Experiment.* Institute of Intergroup Relations, University of Oklahoma, Norman.

Sherif, M. 1966. *In Common Predicament: Social Psychology of Intergroup Conflict and Cooperation.* Addison-Wesley, Reading, MA.

Shonk, J. H. 1992. *Team Based Organizations.* Business One Irwin, Homewood, IL.

Slevin, D. P., and Pinto, J. K. 1988. The project implementation profile: new tool for project managers. *Project Management Journal.*

Smith, C. G. 1979. Age and R&D groups: a reconsideration. *Human Relations* 23:81–93.

Smith, K. K. 1983. An intergroup perspective on individual behavior. In: J. R. Hackman, E. E. Lawler, and L. W. Porter, eds., *Perspective on Behavior in Organizations.* McGraw-Hill, New York, pp. 397–407.

Smith, K. K. 1989. The movement of conflict in organizations: The joint dynamics of splitting and triangulation. *Administrative Science Quarterly* 31:1–20.

Snow, C. C., and Hrebiniak, A. 1980. Strategy, distinctive competence, and organizational performance. *Administrative Science Quarterly* 25.

Snow, C. C., Lipnack, J., and Stamps, J. 1999. The Virtual organization: promises and payoffs, large and small. In: C. L. Cooper and D. M. Rousseau, eds., *The Virtual Organization.* Wiley, New York, pp. 15–30.

Stalk, G., and Hout, T. M. 1990. *Competing Against Time: How Time Based Competition Is Reshaping Global Markets.* New York, The Free Press.

Thompson, J. D. 1967. *Organizations in Action.* McGraw-Hill, New York.

Toney, F. L. 1994. leaders: Actions and traits that result in profitable companies. *Journal of Leadership Studies* 1(4).

Toney, F. L. 1996. A leadership methodology: actions, traits and skills that result in goal achievement. *Journal of Leadership Studies* 3(1).

Toney, F. L., and Powers, R. 1998. *Best Practices of Project Management Groups in Large Functional Organizations.* Project Management Institute, Upper Darby, PA.

Tung, R. L. 1979. Dimensions of organizational environments: an exploratory study of their impact on organizational structure. *Academy of Management Journal* 22:672–693.

Tushman, M., and Nadler, D. A. 1978. An information processing approach to organizational design. *Academy of Management Review* 3(3):613–624.

Tzu, Sun. ca. 500 BC. Reprint by James Clavell, ed., 1983. *The Art of War.* Delacorte Press, New York.

Utterback, J. 1974. Innovation in industry and the diffusion of technology. *Science* 183:620–626.

Utterback, J. M., Allen, T. J., Holloman, J. H., and Sirbu, M. H. 1976. The process of innovation in five industries in Europe and Japan. *IEEE Transactions in Engineering Management* 3–9.

Vesey, J. T. 1991. The new competitors: they think in terms of speed to market. *Academy of Management Executive* 5(2): 23–33.

Von Hippel, E. 1978. A customer active paradigm for industrial project idea generation. *Research Policy* 7:241–266.

Wageman, R., and Baker, G. 1997. Incentives and cooperation: the joint effects of task and reward interdependence on group performance. *Journal of Organizational Behavior* 18.

Weick, K. E. 1979. *The Social Psychology of Organizing.* Addison-Wesley, Reading, MA.

Weick, K. E. 1985. Cosmos vs. chaos: sense and nonsense in electronic contexts. *Organizational Dynamics* Autumn: 50–65.

Weick, K. E. 1993. The collapse of sense making in organizations: the Mann Gulch disaster. *Administrative Science Quarterly* 38:628–652.

Whitener, E., Brodt, S., Korsgaard, M., and Werner, J. 1998. Managers as initiators of trust: an exchange relationship framework for understanding managerial trustworthy behavior. *Academy of Management Review* 23(3).

Whitley, R., and Frost, P. 1973. Task type and information transfer in a government research lab. *Human Relations* 25:537–550.

Womack, J. P., Jones, D. T., and Roos, D. 1990. *The Machine that Changed the World.* Harper Collins, New York.

Zangwill, W. I. 1992. Concurrent engineering: concepts and implementation. *EMR* Winter 1992.

Zangwill, W. I. 1993. *How the World's Best Firms Create New Products.* Lexington Books, New York.

Zirger, B. J., and Maidique, M. 1990. A model of new product development: An empirical test. *Management Science* 36:867–883.

Index